Hydroponic Farming - A Modern Agriculture Technique

Basharat Ali,
Javed Iqbal and Tanveer Ahmed

Published in London, United Kingdom

Hydroponic Farming - A Modern Agriculture Technique
http://dx.doi.org/10.5772/10.5772/intechopen.1003347
Edited by Basharat Ali, Javed Iqbal and Tanveer Ahmed

Contributors
Abate Ayele, Armando Gomez-Vazquez, Bambang Triyatmo, Bonga Lewis Ngcobo, Digafe Alemu, Dongeok Kim, Edmundo Gomez-Mendez, Erlina Ambarwati, Gamachis Korsa, Guangjae Lee, Haifa Ben Hadj Daoud, Hortensia Brito-Vega, Jeongwook Heo, Jorge Pedro Rosa Duarte, Latif Sahubawa, Lemma Tessema, Marco Rosario Butera, Nokuzola Phungula, Pablo García-Lopez, Philile Ngcobo, Setegn Haile, Tesfahun Belay Mihrete, Zandile Maninjwa, Zebenay Dagne

First published in London, United Kingdom, 2025 by IntechOpen
IntechOpen is the global imprint of INTECHOPEN LIMITED, registered in England and Wales, registration number: 11086078, 167-169 Great Portland Street, London, W1W 5PF, United Kingdom

For EU product safety concerns: IN TECH d.o.o., Prolaz Marije Krucifikse Kozulić 3, 51000 Rijeka, Croatia, info@intechopen.com or visit our website at intechopen.com.

British Library Cataloguing-in-Publication Data
A catalogue record for this book is available from the British Library

Hydroponic Farming - A Modern Agriculture Technique
Edited by Basharat Ali, Javed Iqbal and Tanveer Ahmed
p. cm.

This title is part of the Agricultural Sciences Book Series, Volume 20
Topic: Agronomy and Horticulture
Series Editor: W. James Grichar
Topic Editor: Ibrahim Kahramanoglu
Associate Topic Editors: Murat Helvaci and Olga Panfilova

Print ISBN 978-0-85014-835-0
Online ISBN 978-0-85014-834-3
eBook (PDF) ISBN 978-0-85014-836-7
ISSN 3029-052X

If disposing of this product, please recycle the paper responsibly.

IntechOpen Book Series

Agricultural Sciences

Volume 20

Aims and Scope of the Series

The importance of agriculture cannot be overstated. It helps sustain life, as it gives us the food we need to survive and provides opportunities for economic well-being. Agriculture helps people prosper around the world and combines the creativity, imagination, and skill involved in planting crops and raising animals with modern production methods and new technologies. This series includes two main topics: Agronomy and Horticulture, and Animal Farming. This series will help readers better understand the intricacies of production agriculture and provide the new knowledge that is required to be successful. The success of a farmer in modern agriculture requires knowledge of events happening locally as well as globally that impact input decisions and ultimately determine net profit.

Meet the Series Editor

W. James Grichar has been employed with Texas A&M AgriLife Research for over 45 years with an emphasis on research in agronomy, plant pathology, and weed science. He obtained his BS from Texas A&M in 1972 and his Masters of Plant Protection in 1975. He has published 195 journal articles, over 330 research reports and briefs, 11 book chapters, and over 300 abstracts of profession meetings. He also directs research in many crops including corn, grain sorghum, peanuts, and sesame. He has held various positions in different professional societies including the American Peanut Research and Education Society, Southern Weed Science Society, and Texas Plant Protection Conference in addition to being Associate Editor for Peanut Science and Weed Technology. Significant accomplishments have included spearheading efforts to determine the optimum planting time for soybean production along the upper Texas Gulf Coast. These efforts have shown growers that soybean yields can be improved by 10 to 20% by following a late March to early April plant date. He also has been instrumental in developing a herbicide program for peanut production in the south Texas growing region. Through the development and use of herbicides that are effective against major weed problems in the south Texas region, peanut yields have increased by 25 to 30%.

Meet the Volume Editors

Dr. Basharat Ali has worked as an Assistant Professor (Agronomy) in the Department of Agricultural Engineering at Khwaja Fareed University of Engineering and Information Technology in Rahim Yar Khan, Pakistan, since March 2022. Dr. Ali earned his B. Sc (Hons) in Agriculture-Agronomy and M. Sc (Hons) in Agronomy from the University of Agriculture Faisalabad in Pakistan, in 2009 and 2011, respectively, and Ph.D. (Agronomy) from Zhejiang University, Hangzhou, China, in 2014. His specialization includes abiotic stress tolerance, plant nutrition and plant growth regulation. After his Ph.D., Dr. Ali did a postdoc fellowship at Zhejiang University, China, from 2014 to 2016. Then, he did another postdoc fellowship named "Alexander von Humboldt Fellowship (AvH)" from the University of Bonn, Germany, during 2016-2018. Dr Basharat Ali has also served as Assistant Professor (Visiting) at the Department of Agronomy, University of Agriculture Faisalabad, Pakistan, from 2018-2022. Dr Basharat Ali has published over 180 Articles in well-recognized impact factor Journals with an impact factor of 600. Further, Dr. Ali has contributed to several international books, and his publications have gotten around 8,000 citations from other scientists/researchers.

Dr. Javed Iqbal has been working as a Lecturer (Agronomy) in the Department of Agricultural Engineering at Khwaja Fareed University of Engineering and Information Technology in Rahim Yar Khan, Pakistan, since April 2022. He has published over 30 articles in well-recognized impact factor journals with a cumulative impact factor of 65, and his publications got around 400 citations from other scientists/researchers. Further, he has attended and organized several national and international conferences.

Dr. Tanveer Ahmed is an Assistant Professor in the Department of Life Sciences at Khwaja Fareed University of Engineering and Information Technology in Rahim Yar Khan, Pakistan. He completed his Doctorate at the University of Agriculture in Faisalabad, Pakistan, in 2019. After Ph.D. completion, he joined the Department of Zoology at the University of Okara in Pakistan. He served as an Assistant Professor under Pakistan's IPFP program of the Higher Education Commission (HEC). He also served as a Research Associate in the HEC-Pakistan-funded project for almost 3 years. A part of his Ph.D. research work was completed at the Ecological Genetic Research Unit (EGRU) in the Department of Biosciences at the University of Helsinki in Finland. During his Ph.D., his dissertation was titled "Patterns of genetic variability in natural and hatchery populations of Catla catla". His main goal is to achieve excellence through high-quality education and research relevant to local, regional, and national needs, as well as through coordination with leading researchers and educators both in Pakistan and abroad to explore, understand, and conserve animal diversity. He has published over 40 Research Articles in well-recognized impact factor Journals with a cumulative impact factor of 55.

Contents

Preface

Hydroponic farming, a modern and innovative cultivation technique, offers numerous benefits, such as reduced water usage, increased crop yield, and faster growth rates compared to traditional soil-based farming. As the demand for sustainable agricultural practices grows, hydroponics presents an environmentally friendly solution. Plants in hydroponic systems rely entirely on the nutrients dissolved in the water. It is vital to ensure the use of high-quality, balanced nutrient solutions tailored to the specific plants that are being grown. The solution's pH and electrical conductivity (EC) should be regularly monitored to avoid nutrient imbalances. Efficient water circulation and proper oxygenation are essential for maintaining a healthy hydroponic system. Water should be ensured that it constantly moves through the plant's root zones and is adequately aerated to prevent root rot and promote healthy plant development. Indoor hydroponic farms must maintain stable temperature and humidity levels. Most plants thrive in temperatures between 65°F and 75°F (18°C to 24°C). Modern hydroponic farms benefit from technological advancements like sensors and automation. Systems that monitor water levels, nutrient concentrations, and environmental conditions should be utilized. This data enables real-time adjustments and ensures a productive, sustainable farm. Hydroponic systems are versatile and can support various crops, including leafy greens, herbs, fruits, and vegetables. Different crops can be experimented with to optimize yield and cater to diverse market demands.

Basharat Ali, PhD
Assistant Professor (Agronomy),
Department of Agricultural Engineering,
Khwaja Fareed University of Engineering and Information Technology,
Rahim yar Khan, Pakistan

Javed Iqbal, PhD
Lecturer (Agronomy),
Department of Agricultural Engineering,
Khwaja Fareed University of Engineering and Information Technology,
Rahim yar Khan, Pakistan

Tanveer Ahmed, PhD
Assistant Professor,
Department of Life Sciences,
Khwaja Fareed University of Engineering and Information Technology,
Rahim yar Khan, Pakistan

Chapter 1

Crop Substrates for Sustainable Hydroponic Farming

Tesfahun Belay Mihrete

Abstract

Hydroponic farming, as a method of cultivating plants in nutrient-rich water solutions without soil, presents a compelling solution to contemporary food security challenges. This chapter explores the pivotal role of crop substrates in sustainable hydroponic systems, emphasizing their functions in supporting plant growth and their impact on resource efficiency and environmental sustainability. I discuss various types of substrates, including inert materials like perlite and organic alternatives such as coconut coir, focusing on their unique properties and contributions to nutrient management, root health, and water retention. The chapter highlights challenges such as substrate degradation and pH management, alongside opportunities for innovation in substrate technology and regulatory frameworks. It concludes by advocating for the integration of best practices and technological advancements to optimize hydroponic farming for enhanced sustainability, productivity, and resilience in agriculture.

Keywords: hydroponic farming, sustainable substrates, coconut coir, perlite, crop productivity, resource optimization, urban agriculture, environmental sustainability, controlled environment agriculture, soilless cultivation

1. Introduction

Hydroponic farming revolutionizes agriculture by growing plants in nutrient-rich water solutions without using soil [1]. This method has gained prominence as a solution to global food security challenges due to its ability to maximize land use efficiency, conserve water, and minimize environmental impact compared to traditional soil-based farming practices [1, 2]. At the core of successful hydroponics lies the careful selection and management of crop substrates—materials essential for supporting plant roots, ensuring stability, and facilitating nutrient uptake in the absence of soil [3]. The choice of crop substrates in hydroponic systems is pivotal as it profoundly influences key aspects of plant growth such as nutrient availability, root health, and water retention [4]. Farmers utilize various types of substrates in hydroponics, each offering unique advantages and challenges [5]. Inert materials like perlite and rock wool are chemically inert and provide excellent aeration and drainage, crucial for preventing waterlogging and promoting healthy root development [5]. On the other hand, natural materials such as coconut coir and peat moss derive organic substrates, which offer good water-holding capacity and sometimes contribute essential nutrients to plants [6]. However, they may decompose over time, affecting their structural integrity and nutrient retention capabilities [6, 7].

This chapter critically examines the role of crop substrates in sustainable hydroponic farming practices. It delves into the diverse types of substrates used in hydroponic systems, their specific functions in supporting plant growth, and their significant impact on resource efficiency and environmental sustainability. By exploring the intricate relationship between substrate selection and agricultural outcomes, the chapter aims to provide a comprehensive understanding of how farmers, researchers, and policymakers can optimize hydroponic farming practices for long-term sustainability. Key considerations include how different substrate types interact with hydroponic nutrient solutions, influencing factors such as pH levels, nutrient uptake efficiency, and overall crop productivity. The chapter also emphasizes the importance of selecting substrates that balance effective nutrient delivery with sustainable practices, thereby minimizing environmental footprint and resource wastage. Ultimately, by integrating knowledge of substrate characteristics and their effects on plant physiology, hydroponic growers can enhance crop yields, conserve water, and contribute to a more sustainable agricultural system.

2. Types of substrates used in hydroponic systems

2.1 Inert substrates

In hydroponic systems, substrates are essential components that provide physical support to plant roots and facilitate nutrient delivery [8]. One category of substrates commonly used is inert substrates, including materials like perlite, vermiculite, and rock wool as shown in **Figure 1** [9]. These substrates are chemically inert, meaning they do not react with the nutrient solution [9]. They offer excellent aeration and drainage properties, crucial for preventing waterlogging and promoting healthy root development [10]. Their lightweight nature makes them easy to handle, but they necessitate careful management of nutrient levels and pH in the solution since they do not contribute any nutrients themselves [10, 11].

2.2 Organic substrates

Another category comprises organic substrates such as coconut coir and peat moss as shown in **Figure 2** [12]. Natural materials or bio-wastes from humans and animals derive these substrates, as indicated in **Figure 3**, making them more sustainable compared to inert options [12]. They typically possess good water-holding capacity and may contribute some nutrients to the plants [13]. Coconut coir, for example, also exhibits resistance to bacterial and fungal growth [14]. However, organic substrates can decompose over time, affecting their structural integrity and nutrient retention capabilities [15]. This decomposition may require periodic replacement or supplementation with nutrients to maintain optimal growing conditions [15].

2.3 Composite substrates

Composite substrates represent a third category, blending inert and organic materials to capitalize on their combined benefits [16]. These substrates aim to

Figure 1.
Inert substrates in hydroponic farming systems. This figure shows common inert substrates like perlite, vermiculite, pumice, and LECA, used in hydroponic systems for their excellent aeration, drainage, and non-reactive properties, which support healthy root growth and stable nutrient conditions. Source: Olympios [9].

Figure 2.
Organic substrates in hydroponic farming systems. This figure displays organic substrates such as coconut coir and peat moss, valued for their natural water-holding capacity and ability to provide some nutrients, supporting sustainable plant growth in hydroponic systems. Source: Younis et al. [12].

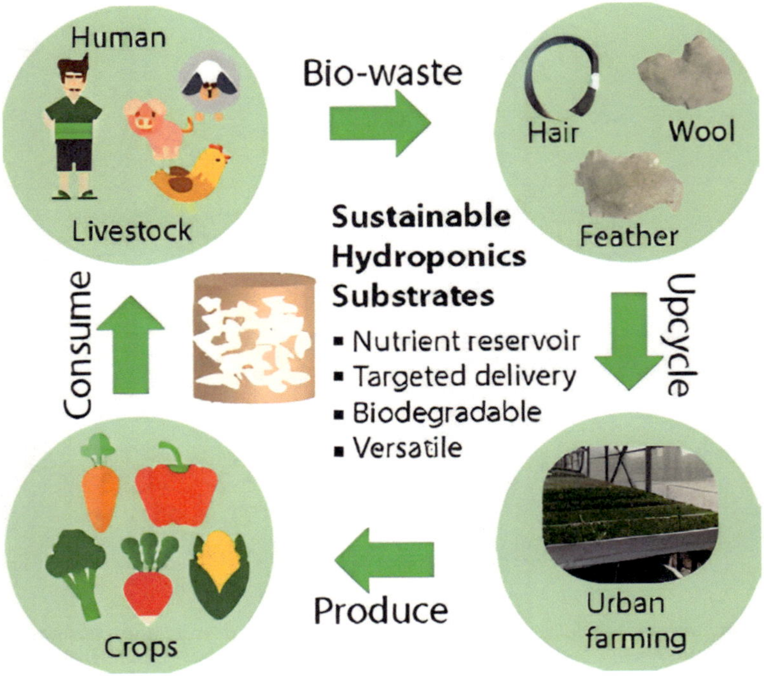

Figure 3.
Cycle and benefits of organic and sustainable substrates in hydroponic farming. This figure illustrates the life cycle of organic substrates, highlighting their sustainability benefits, including resource renewability, biodegradability, and contribution to nutrient cycling within hydroponic farming systems. Source: Younis et al. [12].

enhance nutrient availability and moisture retention while mitigating the individual drawbacks of each substrate type [8]. For instance, mixing perlite with coconut coir, as shown in **Figure 4**, can provide both good aeration and effective nutrient retention [8]. Composite substrates are particularly favored in commercial hydroponic setups where precise control over nutrient delivery and root health is paramount [17].

Figure 4.
Mixing perlite with coconut coir substrate in hydroponic farming. This figure demonstrates the combination of perlite and coconut coir, blending their properties to enhance aeration, drainage, and moisture retention, creating an optimal growing medium for hydroponic plant systems. Source: Zhao et al. [8].

Researchers tailor these substrates to optimize plant growth by providing a balanced environment that supports robust root systems and efficient nutrient uptake [18].

3. Functions of substrates in supporting plant growth

Substrates in hydroponic systems fulfill essential functions that are crucial for supporting plant growth and optimizing crop productivity [19]. One primary function is nutrient delivery. Substrates serve as mediums through which nutrient solutions flow, interacting with plant roots to facilitate nutrient uptake [20]. Growers recognize inert substrates like perlite and rock wool for their excellent drainage and aeration properties [21]. This ensures that roots have access to oxygen while allowing nutrient solutions to permeate freely [19]. Organic substrates such as coconut coir and peat moss not only provide physical support but also possess natural cation exchange capacities [22]. They can retain and release essential nutrients as required by the plants, contributing to efficient nutrient management [23]. Root aeration is another critical function facilitated by substrates [24]. Adequate aeration is essential for root health, allowing for optimal oxygen uptake necessary for root respiration and nutrient absorption [25]. Inert substrates with high porosity ensure good air circulation around the roots, preventing waterlogging and promoting healthy root development [26]. Organic substrates, while generally denser, can still provide adequate aeration when appropriately managed [27]. Techniques such as incorporating perlite or coarse sand into organic substrates can enhance their aeration capacity, further supporting robust root systems and enhancing nutrient uptake efficiency [26, 27].

Managing water availability is equally vital in hydroponic systems, and substrates play a significant role in water retention [28]. Inert substrates like vermiculite and rock wool have high water retention capacities, maintaining consistent moisture levels around the roots [29]. They retain water and nutrients, gradually releasing them as needed by the plants [29]. Organic substrates such as coconut coir also retain water effectively but require careful monitoring to prevent waterlogging or drying out [30]. Effective management of substrate moisture content involves adjusting irrigation schedules, optimizing drainage, and selecting substrates that balance water retention with adequate aeration [31]. This ensures that plants receive sufficient water while minimizing the risk of root diseases and nutrient imbalances [30, 31].

To summarize, substrates in hydroponic systems perform essential functions beyond physical support, regulating nutrient availability, promoting root health through aeration, and managing water retention. By understanding these functions and selecting appropriate substrates, hydroponic growers can optimize growing conditions, enhance crop yields, and maintain overall system efficiency.

4. Impact on resource efficiency and environmental sustainability

Hydroponic farming offers significant advantages in resource efficiency and environmental sustainability, particularly through careful substrate selection and management practices. Firstly, the choice of substrates optimizes resource utilization in hydroponics [32]. Substrates like inert materials (e.g., perlite and rock wool) and organic materials (e.g., coconut coir and peat moss) play crucial roles in efficient water and nutrient management [33]. Inert substrates typically have high water retention and good drainage properties, minimizing water waste through efficient delivery and retention of nutrient solutions to plant roots [34]. This not only reduces water consumption compared to traditional soil-based agriculture but also ensures that plants receive the precise amount of nutrients needed, minimizing nutrient runoff and leaching [35]. Organic substrates, on the other hand, retain nutrients and provide a sustainable alternative to peat moss, which researchers increasingly recognize for its environmental impact due to extraction practices [36]. Another benefit of hydroponics is the efficient use of space, as growers select substrates based on their ability to maximize plant density and growth [37]. Vertical farming systems, for instance, capitalize on substrates that provide structural support while efficiently using vertical space, thereby increasing yield per unit area [38]. Secondly, substrate choices and management practices influence environmental sustainability in hydroponics [39]. Sustainable substrates such as coconut coir and peat moss come from renewable resources and can be recycled or reused, reducing waste and environmental impact compared to traditional soil-based agriculture [40]. Furthermore, substrates like rock wool, despite being inert, can be recycled and reused in hydroponic systems, promoting circular economy principles and reducing overall waste generation [40, 41].

Case studies and research findings highlight the importance of sustainable substrate management in reducing the environmental footprint of hydroponic farming [42]. For instance, optimizing nutrient solutions and irrigation practices based on substrate characteristics can minimize nutrient leaching and runoff, preventing contamination of water bodies and soil [42, 43]. Proper disposal or recycling of substrates at the end of their lifecycle further contributes to environmental sustainability by reducing landfill waste and conserving natural resources [44]. In conclusion, substrate choices and management practices in hydroponic farming have a direct impact on resource utilization efficiency and environmental sustainability. By selecting substrates that enhance water and nutrient efficiency, minimize waste, and support sustainable practices, hydroponic growers can contribute to a more sustainable agricultural system that conserves resources, reduces environmental impact, and enhances food production efficiency.

5. Challenges and opportunities in utilizing crop substrates

5.1 Technical challenges

One of the primary technical challenges in hydroponic farming relates to substrate management. Substrates can degrade over time, affecting their physical structure and nutrient-holding capacity [45]. This degradation can lead to issues such as compaction, reduced aeration, and uneven distribution of water and nutrients to plant roots [45, 46]. Addressing these challenges requires innovative solutions, such as using biodegradable substrates that maintain their integrity over longer periods or developing substrate blends that enhance durability and nutrient retention [47]. P^H fluctuation is another critical technical challenge in hydroponics [48]. Substrates can influence the pH of the nutrient solution, affecting nutrient availability to plants [49]. Managing pH levels involves regular monitoring and adjustment using pH buffers or pH stabilizers to maintain an optimal range for plant growth [50]. Advanced monitoring technologies and automated pH adjustment systems are emerging as solutions to streamline this process and ensure consistent crop performance [51]. Disease management is also a concern in hydroponic systems, where the absence of soil does not eliminate the risk of plant pathogens [52]. Substrates can harbor pathogens or promote their growth under certain conditions [52, 53]. Implementing strict sanitation protocols, using disease-resistant substrates, and integrating biological control methods (e.g., beneficial microbes) are strategies to mitigate disease risks without relying on chemical treatments that could affect environmental sustainability [54].

5.2 Regulatory and economic considerations

Navigating regulatory frameworks is crucial for hydroponic farmers, particularly concerning substrate choices and their impact on food safety and environmental regulations [55]. Regulatory bodies may have specific requirements regarding substrate materials, nutrient management practices, and waste disposal [56]. Compliance with these regulations ensures consumer safety and environmental protection while also influencing the feasibility and scalability of hydroponic operations [57]. Economic factors also play a significant role in substrate utilization in hydroponics [58]. The availability and affordability of substrates can affect production costs and profitability [58, 59]. Market dynamics, such as fluctuations in substrate prices or availability due to supply chain disruptions, influence farmers' decisions on substrate selection and management practices [60]. Economic analyses help in evaluating the cost-effectiveness of different substrate options and optimizing resource allocation to maximize returns on investment [61]. In the final analysis, while hydroponic farming offers numerous benefits in terms of resource efficiency and crop productivity, it also presents challenges that require careful consideration and innovative solutions. Addressing technical challenges related to substrate degradation, pH management, and disease control is essential for sustainable farming practices. Navigating regulatory frameworks and economic considerations ensures compliance with standards and optimizes operational efficiency in hydroponic agriculture. By overcoming these challenges and seizing opportunities through technological advancements and strategic decision-making, hydroponic growers can enhance sustainability, profitability, and resilience in modern agriculture.

6. Best practices and innovative solutions for sustainable hydroponic farming

In the realm of sustainable agriculture, particularly within hydroponics, the implementation of best practices is essential for optimizing resource use and environmental impact [62]. One critical area involves nutrient management, where precision delivery systems in hydroponics can minimize waste and maximize plant uptake [63]. For instance, techniques such as drip irrigation tailored to crop needs based on growth stages ensure efficient nutrient utilization [63, 64]. This approach not only enhances plant health but also reduces the overall environmental footprint associated with nutrient runoff [65].

Water conservation is another pivotal aspect of sustainable hydroponic farming [66]. Closed-loop systems that recycle water and nutrients significantly decrease water usage compared to conventional soil-based agriculture [67]. Such systems demonstrate a commitment to sustainability by minimizing water depletion and nutrient pollution in surrounding ecosystems [68]. Additionally, integrating renewable energy sources, such as solar panels, to power hydroponic facilities underscores a broader strategy for energy efficiency and reducing carbon emissions [69]. Innovative solutions further propel sustainable agriculture forward by leveraging cutting-edge technologies [70]. For instance, advancements in substrate technologies, such as biodegradable hydrogels or aeroponic systems, offer promising avenues for improving resource efficiency and crop yields [71]. These technologies not only reduce water and nutrient requirements but also enhance plant growth in controlled environments [72]. Moreover, the integration of Internet of Things (IoT) sensors and data analytics enables real-time monitoring of plant health and environmental conditions, optimizing nutrient delivery and predicting crop yields with greater precision as indicated in **Figure 5** [73].

Real-world applications in countries like the Netherlands and Singapore display successful implementations of advanced hydroponic systems [74]. In the Netherlands, high-tech greenhouse facilities employ sophisticated climate control and water management technologies for year-round vegetable production [75]. Similarly, Singapore's vertical farming initiatives in urban spaces utilize hydroponic systems to maximize land productivity and reduce dependence on food imports [76]. Looking ahead, future research should focus on developing standardized metrics to assess sustainability in hydroponic farming comprehensively. Exploring new substrates and scaling up successful innovations will also be critical to advancing sustainable agriculture and addressing global food security challenges in the years to come. By integrating best practices, innovative technologies, and real-world case studies, hydroponic agriculture can continue to evolve as a sustainable solution for meeting growing food demand while preserving natural resources.

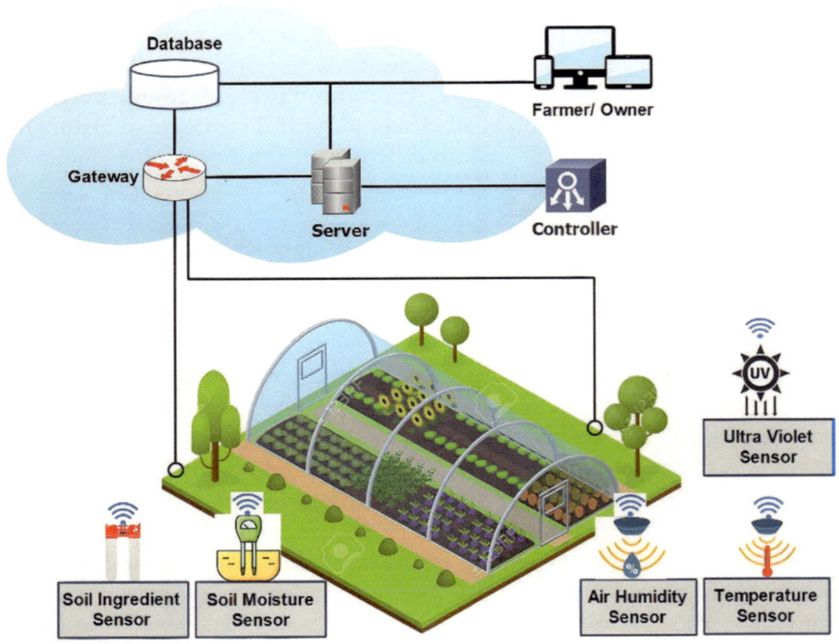

Figure 5.
Internet of Things (IoT) sensors in smart farming. This figure illustrates the use of IoT sensors in smart farming, displaying their role in real-time monitoring of environmental conditions, optimizing nutrient delivery, and improving precision in crop management within hydroponic systems. Source: Alahmad et al. [73].

7. Conclusion

In conclusion, the effective selection and management of crop substrates are fundamental to advancing sustainable hydroponic farming practices. This chapter has underscored how substrates not only provide physical support and facilitate nutrient delivery but also play a critical role in optimizing resource efficiency and minimizing environmental impact. From inert materials like perlite and rock wool to organic options such as coconut coir and peat moss, growers must carefully consider the distinct benefits and challenges of each substrate type to enhance crop productivity and sustainability. By understanding the functions of substrates in supporting plant growth—such as nutrient management, root aeration, and water retention—hydroponic growers can tailor their systems to maximize efficiency while minimizing water use and nutrient runoff. Moreover, the integration of innovative technologies and best practices, such as precision nutrient delivery systems and closed-loop water recycling, further enhances sustainability efforts in hydroponic agriculture. Looking forward, ongoing research and practical applications will continue to refine substrate choices, improve cultivation techniques, and expand the scalability of hydroponic farming. By embracing these advancements and fostering collaboration among farmers, researchers, and policymakers, hydroponic agriculture can contribute significantly to global food security goals while promoting environmental stewardship and resource conservation.

Acknowledgements

The author acknowledges the use of Grammarly for language polishing of the manuscript.

Author details

Tesfahun Belay Mihrete
Department of Horticulture, College of Agriculture and Environmental Sciences, Bahir Dar University, Bahir Dar, Ethiopia

*Address all correspondence to: tesfahunbelay2010@gmail.com

References

[1] Kaur S, Dewan B. Hydroponics: A review on revolutionary technology for sustainable agriculture. Asian Journal of Agricultural and Horticultural Research. 2023;**10**(4):297-302

[2] Shah P. Hydroponic—A new era for agriculture? Webology. 2021;**18**(6):121-140. ISSN: 1735-188X

[3] Arumugam T, Sandeep G, Maheswari MU. Soilless farming of vegetable crops: An overview. The Pharma Innovation Journal. 2021;**10**(1):773-785

[4] Sharma A, Manpoong C, Devadas VS, Kartha BD, Pandey H, Wangsu M. Crop hydroponics, phyto-hydroponics, crop production, and factors affecting soilless culture. ACS Agricultural Science & Technology. 2022;**2**(6):1134-1150

[5] Patil ST, Kadam US, Mane MS, Mahale DM, Dhekale JS. Hydroponic growth media (substrate): A review. International Research Journal of Pure and Applied Chemistry. 2020;**21**(23):106-113

[6] Atzori G, Pane C, Zaccardelli M, Cacini S, Massa D. The role of peat-free organic substrates in the sustainable management of soilless cultivations. Agronomy. 2021;**11**(6):1236

[7] Gruda NS, Bragg N. Developments in alternative organic materials for growing media in soilless culture systems. In: Advances in Horticultural Soilless Culture. Vol. 6(2). Cambridge, UK: Burleigh Dodds Science Publishing; 2021. pp. 73-106

[8] Zhao Z, Xu T, Pan X, Susanti, White JC, Hu X, et al. Sustainable nutrient substrates for enhanced seedling development in hydroponics. ACS Sustainable Chemistry and Engineering. 2022;**10**(26):8506-8516

[9] Olympios CM. Soilless media under protected cultivation rock wool, peat, perlite and other substrates. In: Symposium on Soil and Soilless Media under Protected Cultivation in Mild Winter Climates 323. Vol. 10(2). 1992. pp. 215-234

[10] Sonneveld C, Voogt W, Sonneveld C, Voogt W. Substrates: Chemical characteristics and preparation. In: Plant Nutrition of Greenhouse Crops. Vol. 11(3). Delhi, India: International Journal of Research in Agronomy, Integrated Publishers; 2009. pp. 227-256

[11] Kang JY, Lee HH, Kim KH. Physical and chemical properties of inorganic horticultural substrates used in Korea. In: International Symposium on Growing Media and Hydroponics 644. Vol. 6(2). 2001. pp. 237-241

[12] Younis A, Ahsan M, Akram A, Lim KB, Zulfiqar F, Tariq U. Use of organic substrates in sustainable horticulture. In: Hasanuzzaman M, Hawrylak Nowak B, Islam T, Fujita M, editors. Bio Stimulants for Crop Production and Sustainable Agriculture. Vol. 8(2). Wallingford, UK: Centre for Agriculture and Biosciences International (CABI); 2022. pp. 122-138

[13] Thomas GV, Krishnakumar V, Dhanapal R, Srinivasa Reddy DV. Agro-management practices for sustainable coconut production. In: The Coconut Palm (*Cocos nucifera* L.)-Research and Development Perspectives. Vol. 5(1). Singapore: Springer; 2018. pp. 227-322

[14] Subramanian P, Gupta A, Gopal M, Selvamani V, Mathew J, Surekha IS.

Coconut (*Cocos nucifera* L.). In: Soil Health Management for Plantation Crops: Recent Advances and New Paradigms. Vol. 4(2). Singapore: Springer Nature Singapore; 2024. pp. 37-109

[15] Ferreira V, Elosegi A, Tiegs SD, von Schiller D, Young R. Organic matter decomposition and ecosystem metabolism as tools to assess the functional integrity of streams and rivers—A systematic review. Water. 2020;**12**(12):3523

[16] Arif ZU, Khalid MY, Sheikh MF, Zolfagharian A, Bodaghi M. Biopolymeric sustainable materials and their emerging applications. Journal of Environmental Chemical Engineering. 2022;**10**(4):108159

[17] Rajendran S, Domalachenpa T, Arora H, Li P, Sharma A, Rajauria G. Hydroponics: Exploring innovative sustainable technologies and applications across crop production, with emphasis on potato mini-tuber cultivation. Heliyon. 2024;**7**(2):177-181

[18] Kumar S, Singh M, Yadav KK, Singh PK. Opportunities and constraints in hydroponic crop production systems: A review. Environment Conservation Journal. 2021;**22**(3):401-408

[19] Al Meselmani MA. Nutrient solution for hydroponics. In: Recent Research and Advances in Soilless Culture. Vol. 3(1). London, UK: IntechOpen; 2022. pp. 121-131

[20] Wang H, Inukai Y, Yamauchi A. Root development and nutrient uptake. Critical Reviews in Plant Sciences. 2006;**25**(3):279-301

[21] Othman Y, Bataineh K, Al-Ajlouni M, Alsmairat N, Ayad J, Shiyab S, et al. Soilless culture: Management of growing substrate, water, nutrient, salinity, microorganism and product quality. Fresenius Environmental Bulletin. 2019;**28**(4):3249-3260

[22] Abad M, Noguera P, Puchades R, Maquieira A, Noguera V. Physico-chemical and chemical properties of some coconut coir dusts for use as a peat substitute for containerized ornamental plants. Bioresource Technology. 2002;**82**(3):241-245

[23] Carlile WR, Cattivello C, Zaccheo P. Organic growing media: Constituents and properties. Vadose Zone Journal. 2015;**14**(6):1-3

[24] Drew MC. Soil aeration and plant root metabolism. Soil Science. 1992;**154**(4):259-268

[25] Ben-Noah I, Friedman SP. Review and evaluation of root respiration and of natural and agricultural processes of soil aeration. Vadose Zone Journal. 2018;**17**(1):1-47

[26] De Castro J, Hill RD, Stasolla C, Badea A. Waterlogging stress physiology in barley. Agronomy. 2022;**12**(4):780

[27] Chan YJ, Chong MF, Law CL, Hassell DG. A review on anaerobic–aerobic treatment of industrial and municipal wastewater. Chemical Engineering Journal. 2009;**155**(1-2):1-8

[28] Gruda NS. Increasing sustainability of growing media constituents and stand-alone substrates in soilless culture systems. Agronomy. 2019;**9**(6):298

[29] Morgan L. Substrate-based Hydroponic Systems. Vol. 4(2). Wallingford, UK: CABI Publisher; 2021. pp. 77-99

[30] Carlile WR, Raviv M, Prasad M. Organic soilless media components. In: Soilless Culture. Vol. 6(1). Cambridge,

Massachusetts, USA: Academic Press; 2019. pp. 303-378

[31] Fussy A, Papenbrock J. An overview of soil and soilless cultivation techniques—Chances, challenges and the neglected question of sustainability. Plants. 2022;**11**(9):1153

[32] Farvardin M, Taki M, Gorjian S, Shabani E, Sosa-Savedra JC. Assessing the physical and environmental aspects of greenhouse cultivation: A comprehensive review of conventional and hydroponic methods. Sustainability. 2024;**16**(3):1273

[33] Bar-Tal A, Saha UK, Raviv M, Tuller M. Inorganic and synthetic organic components of soilless culture and potting mixtures. In: Soilless Culture. Vol. 7(1). Cambridge, UK: Elsevier; 2019. pp. 259-301

[34] Massa D, Magán JJ, Montesano FF, Tzortzakis N. Minimizing water and nutrient losses from soilless cropping in southern Europe. Agricultural Water Management. 2020;**241**(5):106395

[35] Goulding K, Jarvis S, Whitmore A. Optimizing nutrient management for farm systems. Philosophical Transactions of the Royal Society, B: Biological Sciences. 2008;**363**(1491):667-680

[36] Pascual JA, Ceglie F, Tuzel Y, Koller M, Koren A, Hitchings R, et al. Organic substrate for transplant production in organic nurseries. A review. Agronomy for Sustainable Development. 2018;**9**(2):1-23

[37] Gómez C, Currey CJ, Dickson RW, Kim HJ, Hernández R, Sabeh NC, et al. Controlled environment food production for urban agriculture. HortScience. 2019;**54**(9):1448-1458

[38] Maurya PK, Karde RY, Bayskar AS, Charitha N. Innovative technologics such as vertical farming and hydroponics to grow crops in controlled environments. Advanced Farming Technology. 2023;**10**(5):84-90

[39] Szekely I, Jijakli MH. Bioponics as a promising approach to sustainable agriculture: A review of the main methods for producing organic nutrient solution for hydroponics. Water. 2022;**14**(23):3975

[40] Rai A, Smriti S, Kaushal S. Utilising crop residues as hydroponic media for sustainable food production system. International Journal of Research in Agronomy. 2024;**7**(4):73-78

[41] Manríquez-Altamirano A, Sierra-Pérez J, Muñoz P, Gabarrell X. Analysis of urban agriculture solid waste in the frame of circular economy: Case study of tomato crop in integrated rooftop greenhouse. Science of the Total Environment. 2020;**734**(10):139375

[42] Vinci G, Rapa M. Hydroponic cultivation: Life cycle assessment of substrate choice. British Food Journal. 2019;**121**(8):1801-1812

[43] Bilderback TE. Water management is key in reducing nutrient runoff from container nurseries. HortTechnology. 2002;**12**(4):541-544

[44] JeyaSundar PG, Ali A, Zhang Z. Waste treatment approaches for environmental sustainability. In: Microorganisms for Sustainable Environment and Health. Vol. 10(2). Cambridge, UK: Elsevier; 2020. pp. 119-135

[45] Olubanjo OO, Alade AE. Growth and yield response of tomato plants grown under different substrates culture. Journal of Sustainable Technology. 2018;**9**(2):110-123

[46] Bihari C, Ahamad S, Kumar M, Kumar A, Kamboj AD, Singh S, et al.

Innovative soilless culture techniques for horticultural crops: A comprehensive review. International Journal of Environment and Climate Change. 2023;**13**(10):4071-4084

[47] Dalton B, Bhagabati P, De Micco J, Padamati RB, O'Connor K. A review on biological synthesis of the biodegradable polymers polyhydroxyalkanoates and the development of multiple applications. Catalysts. 2022;**12**(3):319

[48] Sambo P, Nicoletto C, Giro A, Pii Y, Valentinuzzi F, Mimmo T, et al. Hydroponic solutions for soilless production systems: Issues and opportunities in a smart agriculture perspective. Frontiers in Plant Science. 2019;**10**(3):923

[49] Trejo-Téllez LI, Gómez-Merino FC. Nutrient solutions for hydroponic systems. In: Hydroponics—A Standard Methodology for Plant Biological Researches. Vol. 21(3). Rijeka, Croatia: InTech; 2012. pp. 1-22

[50] Felle HH. pH regulation in anoxic plants. Annals of Botany. 2005;**96**(4):519-532

[51] Yanes AR, Martinez P, Ahmad R. Towards automated aquaponics: A review on monitoring, IoT, and smart systems. Journal of Cleaner Production. 2020;**263**(10):121571

[52] Mihajlović M, Rekanović E, Hrustić J, Grahovac M, Tanović B. Methods for management of soilborne plant pathogens. Pesticidi i Fitomedicina. 2017;**32**(1):9-24

[53] Suárez-Cáceres GP, Pérez-Urrestarazu L, Avilés M, Borrero C, Eguíbar JR, Fernández-Cabanás VM. Susceptibility to water-borne plant diseases of hydroponic vs. aquaponics systems. Aquaculture. 2021;**544**(11):737093

[54] Collinge DB, Jensen DF, Rabiey M, Sarrocco S, Shaw MW, Shaw RH. Biological control of plant diseases—What has been achieved and what is the direction? Plant Pathology. 2022;**71**(5):1024-1047

[55] Fuentes-Peñailillo F, Gutter K, Vega R, Silva GC. New generation sustainable technologies for soilless vegetable production. Horticulturae. 2024;**10**(1):49

[56] Rosemarin A, Macura B, Carolus J, Barquet K, Ek F, Järnberg L, et al. Circular nutrient solutions for agriculture and wastewater—A review of technologies and practices. Current Opinion in Environmental Sustainability. 2020;**45**(5):78-91

[57] Juarez MI. A feasibility study of hydroponic shipping container farms in businesses and schools: Identifying the influential factors, benefits, and challenges [Master of Education thesis]. Texas, United States of America: Texas State University; 2018

[58] Hameed MK, Umar W, Razzaq A, Aziz T, Maqsood MA, Wei S, et al. Differential metabolic responses of lettuce grown in soil, substrate and hydroponic cultivation systems under NH4+/NO3− application. Metabolites. 2022;**12**(5):444

[59] Sathyanarayana SR, Gangadhar WV, Badrinath MG, Ravindra RM, Shriramrao AU. Hydroponics: An intensified agriculture practice to improve food production. Reviews in Agricultural Science. 2022;**10**(3):101-114

[60] Csambalik L, Divéky-Ertsey A, Gál I, Madaras K, Sipos L, Székely G, et al. Sustainability perspectives of organic farming and plant factory systems—From divergences towards synergies. Horticulturae. 2023;**9**(8):895

[61] Pattillo DA. Needs assessment and practical solutions for the aquaponics industry [doctoral dissertation]. Auburn, Alabama, USA: Auburn University

[62] Dorr E, Goldstein B, Horvath A, Aubry C, Gabrielle B. Environmental impacts and resource use of urban agriculture: A systematic review and meta-analysis. Environmental Research Letters. 2021;**16**(9):093002

[63] Bamsey M, Graham T, Thompson C, Berinstain A, Scott A, Dixon M. Ion-specific nutrient management in closed systems: The necessity for ion-selective sensors in terrestrial and space-based agriculture and water management systems. Sensors. 2012;**12**(10):13349-13392

[64] Rufí-Salís M, Petit-Boix A, Villalba G, Sanjuan-Delmás D, Parada F, Ercilla-Montserrat M, et al. Recirculating water and nutrients in urban agriculture: An opportunity towards environmental sustainability and water use efficiency? Journal of Cleaner Production. 2020;**261**(10):121213

[65] Brar AS, Kaur K, Sindhu VK, Tsolakis N, Srai JS. Sustainable water use through multiple cropping systems and precision irrigation. Journal of Cleaner Production. 2022;**333**(10):130117

[66] Mielcarek A, Kłobukowska K, RodziewiczJ,JanczukowiczW,BryszewskiKŁ. Water nutrient management in soilless plant cultivation versus sustainability. Sustainability. 2023;**16**(1):152

[67] Ragany M. Closed-loop greenhouse agriculture systems [doctoral dissertation]. Ontario, Canada: McMaster University;

[68] Richter BD, Mathews R, Harrison DL, Wigington R. Ecologically sustainable water management: Managing river flows for ecological integrity. Ecological Applications. 2003;**13**(1):206-224

[69] Lachheb A, Marouani R, Mahamat C, Skouri S, Bouadila S. Fostering sustainability through the integration of renewable energy in an agricultural hydroponic greenhouse. Engineering, Technology & Applied Science Research (ETASR). 2024;**14**(2):13398-13407

[70] Júnior MR, de Almeida Moreira BR, dos Santos Carreira V, de Brito Filho AL, Trentin C, de Souza FL, et al. Precision agriculture in the United States: A comprehensive meta-review inspiring further research, innovation, and adoption. Computers and Electronics in Agriculture. 2024;**221**:108993

[71] Kalossaka LM, Sena G, Barter LM, Myant C. 3D printing hydrogels for the fabrication of soilless cultivation substrates. Applied Materials Today. 2021;**24**(2):101088

[72] Lakhiar IA, Gao J, Syed TN, Chandio FA, Buttar NA. Modern plant cultivation technologies in agriculture under controlled environment: A review on aeroponics. Journal of Plant Interactions. 2018;**13**(1):338-352

[73] Alahmad T, Neményi M, Nyéki A. Applying IoT sensors and big data to improve precision crop production: A review. Agronomy. 2023;**13**(10):2603

[74] Halgamuge MN, Bojovschi A, Fisher PM, Le TC, Adeloju S, Murphy S. Internet of Things and autonomous control for vertical cultivation walls towards smart food growing: A review. Urban Forestry & Urban Greening. 2021;**61**(5):127094

[75] Marcelis LF, Costa JM, Heuvelink E. Achieving sustainable greenhouse

production: Present status, recent advances and future developments. In: Achieving Sustainable Greenhouse Cultivation. Vol. 10(2). Cambridge, United Kingdom: Burleigh Dodds Science Publishing; 2019. pp. 1-4

[76] Al-Kodmany K. The vertical farm: A review of developments and implications for the vertical city. Buildings. 2018;**8**(2):24

Chapter 2

Perspective Chapter: Biochar as a Hydroponic Substrate in *Capsicum chinense* Cultivation in the Humide Tropic

Edmundo Gomez-Mendez, Pablo García-López, Hortensia Brito-Vega and Armando Gomez-Vazquez

Abstract

In Mexico, the habanero chili (*Capsicum chinense*) is one of the most economically important vegetables. The objective of this chapter was to evaluate the production of the habanero chili (*C. chinense*) crop on biochar substrate in a hydroponic system under a protected structure. The chili variety that was working with was Orange. The experiment was established with a completely randomized design with five treatments and 10 repetitions, totaling 50 experimental units. Three mixtures of biochar and tepetzil were evaluated, in addition to a treatment with 100% biochar and a control treatment (T1) with only tepetzil. The data were analyzed by analysis of variance and the Tukey test ($P \leq 0.05$) using the Statistical Analysis System statistical program. The results show no differences in terms of fruit weight; however, there was a significant effect on the yield of the habanero pepper. The results demonstrate a clear effect of biochar and hydroponics on yield, where 50% was used; the highest value was obtained with 56.30 T ha-1. However, tropical agriculture usually plays a very important role in the country's economy. Therefore, the implementation of hydroponics as an alternative form of crop production helps promote environmental protection as well as sustainability.

Keywords: tropical vegetables, habanero chili, technology, tepetzil, crops

1. Introduction

Habanero peppers are predominantly grown in four main regions in Mexico at a national level: the Pacific, North-Central, Gulf of Mexico and Yucatan Peninsula. However, the states in the southeast, namely Yucatan, Tabasco, Campeche and Quintana Roo, contribute the most to production, with 39, 30, 8 and 6%, respectively. Even with Chiapas contributing 3%, the southeastern states remain the top producers [1, 2]. The cultivation of habanero chili in controlled environments has led to an increase in the use of different substrates such as tezontle, perlite, sand, rock wool, coconut fibers, tepetzil or tepojal, peat, vermiculite and pumacite. The issue occurs when these components are hard to reach because of their distance, availability, or cost, so it is crucial to find substitutes that are nearby and make use of waste or by-products that can serve the same purpose as a commercial base [3, 4]. Several advanced and industrialized nations, such as the United States, the United Kingdom, Spain, Japan and the Netherlands, have seen a rise in the utilization of soilless indoor farming through hydroponic systems. This is performed in return for soil, inorganic materials, or utilizing organic waste to boost vegetable production and address the challenges presented by climate change [5]. Hydroponic is changing the way agricultural crops are grown globally because of its low environmental impact, enhanced pest management and increased crop output. Alternatively, hydroponics, known as soilless agriculture, is seen to mitigate soil-based agriculture's susceptibility to environmental changes [6]. Additionally, there is a growing need for food production in city settings where there is not much land available for sustainable agriculture, particularly in reducing the carbon emissions from food supply and transportation [7, 8]. Hydroponic nutrient solutions help minimize water and nutrient waste by collecting and recycling drainage, typically utilizing just 5% of the water required in traditional agriculture for the same level of production [8]. This raises the possibility of using charcoal residues as biochar to replace various substrates. Although the use of charcoal in agriculture as a soil amendment or soil improver has been known for thousands of years [9–11], its use as a substrate or soilless growing medium is relatively recent and there has been little focus on charcoal as a substrate or biochar in soilless crops [12, 13]. Fortunately, more information is now available on the subject and there is even a compilation of research on the use of charcoal and other raw materials as biochar in agriculture and as a medium for crops or substrates [14]. Biochar can be used as a substrate to partially replace the commercial substrates used today; However, the characterization of these materials is a practice that has not been given the importance it deserves, since it plays an essential role in the agronomic management of crops [15–18], allows knowing their characteristics and identifying which substrates can be more optimal according to the requirements of the crop in question. The most important quality of hydroponics is that it does not need the soil at any stage of the plant's development, neither as a support or as a source of nutrients, since the water contains dissolved nutrients from which the plant absorbs them directly, achieving greater efficiency in the regulation of nutrition and fertilizer use [4]. The objective of this work is to learn about the cultivation of habanero peppers under the hydroponic system and as substrate biochar and tepetzil (v/v) to better understand the possible effects on *Capsicum* sp.

2. Habanero chili (*C. chinense*)

In Mexico, the species within the genus *Capsicum* are referred to as "chili, a word that comes from the Nahuatl word chili (**Figure 1**) [19, 20]. This genus has about 38 species, of which 12 (including varieties) are used by man; only five species have been domesticated and cultivated (*Capsicum annuum* L., *C. chinense* Jacq., *C. pubescens* L., *C. frutescens* L. and *C. baccatum* L.), the rest are wild and semi-cultivated species [21, 22].

Nowadays, the cultivation of habanero peppers is considered a traditional crop in southeastern Mexico, forming part of the regional gastronomy, representing a great opportunity for producers due to the high demand for habanero marketed both as fresh fruit and for the preparation of sauces, pastes and dehydrated products, which

Figure 1.
"Chile habanero" habanero chili in hydroponic system with protected structure.

are used to accompany our food [23–25]. In addition, the Yucatan Peninsula area (Yucatan, Quintana Roo and Campeche) has had the Denomination of Origin for habanero peppers since 2010 [26]. According to the Ministry of Economy [26], it covers the finished product or products, including their characteristics, components, form of extraction and production or elaboration processes (**Figure 2**). However, this does not minimize the importance that the state of Tabasco represents with respect to the production of habanero peppers. At the national level, habanero peppers are

Figure 2.
Habanero chili on carbon substrate in hydroponic system.

mainly produced in four regions of the country: the Pacific, north-central, Gulf of Mexico and Yucatan Peninsula; however, the states located in the southeast of Mexico are the ones that have the largest share with respect to production, which are Yucatan, Tabasco, Campeche and Quintana Roo, which contribute 39, 30, 8 and 6%, respectively [6, 27]. Of the three varieties of habanero peppers for the tropics of Mexico (Uxmal, Jaguar, Mayapán and Calakmul), it is known that the varieties Mayapán and Calakmul adapt well in the state of Tabasco [28, 29], in addition, that there are studies where Longoria [30] and Orange [31] varieties have been used in the Tabasco region.

3. Biochar and tepetzil

Biochar is a fuel resulting from the combustion at high temperatures and without oxygen (or very low concentrations) of wood for a certain time, where it undergoes a decomposition of lignin, cellulose, hemicellulose and loss of extractables, causing it to be a solid, fragile and porous material [32, 33]. This process is called pyrolysis, where to obtain charcoal, temperatures of 300–700°C can be handled according to the different processes for obtaining it: torrefaction, traditional slow pyrolysis, modern slow pyrolysis, fast pyrolysis, gasification or hydrothermal carbonization [34, 35], the parameters that intervene in the properties of charcoal are: (1) the type of organic matter used; (2) the carbonization environment (temperature and aeration) and (3) the additions during the carbonization process (**Figure 3**). Biochar made from agricultural waste is an alternative substrate for hydroponic production. The literature suggests that there is better absorption of several nutrients in the presence of biochar; However, the results seem to be dependent on the physical and structural characteristics of the carbons, so it is not possible to generalize the results [35]. As it is a recalcitrant form of carbon, it acts as a long-term reservoir of this element, delaying its return to the atmosphere as CO_2, a situation that contributes to mitigating climate change. An additional benefit is that the raw material for its production comes from

Figure 3.
Substrate with (a) charcoal 100%, (b) tepetzil 100% and (c) charcoal 50 and tepetzil 50%, respectively, in the production of habanero chili for a hydroponic system.

organic waste, which often causes environmental pollution problems [36]. Nowadays, charcoal has several uses, highlighting its use as energy, caloric, for composting, in artistic and food areas [37], in addition to its use as a pigment in the food industry, in steel production, to purify water and in the medical industry. Its use is increasingly expanding to diverse areas or industries; one of these areas is agriculture, both in conventional agriculture as a soil improver or restorer, amendments and/or fertilizer [38], as in soilless crop production where it has been tested as a substrate in hydroponic systems [9, 39, 40]. Accordingly, when charcoal is used in agriculture, it is often referred to as biochar; however, this term does not refer only to charcoal. According to the International Biochar Initiative [41], biochar is a solid material obtained from the thermochemical conversion of biomass carbonization in an oxygen-limited environment, a carbonized material that has undergone a pyrolysis process with low or no oxygen conditions similar to charcoal. However, the difference lies in two particularities: (1) to produce biochar, a great variety of organic materials or feedstocks can be used [42, 43] such as plant residues, wood or tree bark [44] and biosolids [45], among others and (2) charcoal or any other carbonized material that is intended for agricultural use is referred to as biochar [46], either for agricultural soils or as a component of substrate for crops; Products similar to biochar may not be called this way if their use is directed to the energy industry, as fuel, in filtering processes, as reducers in the metallurgical industry [47].

As already mentioned, in hydroponics, it is very usual to see the use of substrates as a growing medium, the most frequent being tezontle, perlite, sand, rock wool, coconut fiber, tepetzil or tepojal, peat, vermiculite and pumacite [48]. In Mexico, coconut dust or fiber, tezontle, perlite, pumacite, tepetzil, compost, cucumber bark peat, rice husks and worm humus, among other materials, are used as substrates for various crops [49].

Tepetzil, also known as tepojal, pumita or caltete, is an inorganic material that comes from volcanic foam when lava with high water content and volatile gases is expelled from the volcano. Its use as a substrate is because it improves drainage in soils and in mixture with other substrates, presents high porosity and moisture retention, is light and has neutral pH. However, it is a component that is not recommended for use on its own [50]. Furthermore, unlike tezontle, tepetzil is even more limited in terms of its availability; in Mexico, there are two tepetzil accumulation sites: one in Calimaya, State of Mexico, near Nevado de Toluca and another in Perote, Veracruz, near the Orizaba peak volcano [3]. The same author indicates that the long-term availability of tepojal is uncertain due to the great demand by manufacturers of light blocks for construction and mentions that the tepojal from Calimaya is of lower quality and is increasingly mixed with soil particles.

3.1 Characteristics physicochemical properties of a biochar and tepetzil

The substrate should be light, hold seedlings or cuttings firmly, retain sufficient moisture, be sufficiently porous to drain excess water, allow root aeration, be free of weeds, nematodes and diseases, be suitable for sterilization without changing its properties, have sufficient nutrients for initial healthy plant development, low salinity, adequate pH, be stable and not swell or shrink excessively or form crusts in the sun [51], the substrate must have a high capacity to retain readily available water, high aeration, low bulk density, high porosity, low salinity, high buffer capacity, low decomposition speed, structural stability, reproducibility and availability, low cost and easy handling (mixing and disinfection) [52].

No matter what raw material is used as substrate, it is important that these have certain characteristics which are segmented into physical characteristics, chemical characteristics and biological characteristics (the latter in the case of organic materials); within the physical characteristics, we have the real and apparent density, granulometric distribution, porosity and aeration, water retention, permeability, pore size distribution, structural stability; chemical characteristics such as cation exchange capacity, pH, electrical conductivity, buffer capacity, nutrient content, C/N ratio; and biological characteristics such as microbial population, organic matter content, state and speed of decomposition [53, 54]. **Table 1** presents some optimal values for some of the physical and physicochemical properties in the substrates.

Since the characteristics present in biochar (and any other biochar) are dependent on the raw material and type of pyrolysis used, it is necessary to evaluate its characteristics and its incorporation as a hydroponic substrate to understand and explain the effects on the different plants produced in these soilless growing media [55–57]. However, the characterization of substrates is a process that has not been given much importance and yet plays an essential role in the agronomic management of crops [58].

The need to find partial or complete substitutes not only for tepetzil but also for any commercial substrate is important; new substrates must be found that are even considered by-products or residues from other industries to reduce their

Properties	**Unit**	**Optimum value**
Particle size	mm	0.25–2.50
Bulk density • In open air container • In greenhouse	g•cm^3	0.50–0.75 0.15
Real density • Organic materials • Mineral materials	g•cm^3	1.45 2.65
Total pore space	% vol.	>85
Aeration capacity	% vol.	10–30
Readily available water	% vol.	20–30
Reserve water	% vol.	4–10
Total available water	% vol.	24–40
Concentration		<30
pH (pH in H_2O) • For cultural media • For organic materials		5.5–6.5 5.2–5.5
Electrical conductivity (1:5-v:v)	mS/cm	0.36–0.65
Cation exchange capacity • Permanent fertigation • Occasional fertigation	meq•100 g	Nil or very low >20

Table 1.
Optimum levels of physical and physicochemical properties of a biochar and tepetzil.

environmental impact by not giving them a productive use, for example, due to the hardiness of the biochar pits in Tabasco, charcoal residue is produced that is not marketed due to its reduced size [59].

4. Hydroponic

Hydroponic, from the Greek "hydro" (water) and "ponos" (labor or work), is a soilless crop production technique [60] that has been practiced for centuries, although it has been used commercially for the last 50 years and has been increasing, contributing to sustainable agriculture as a new technology [61]. It is a technique that is not specific to any type of vegetable, so leaf vegetables, stem vegetables, inflorescences, roots, pods, fruits and bulbs can be grown, thus being able to cultivate from home cultivation for own consumption, to important productions for commercialization [5].

The most important quality of hydroponics is that it does not need the soil at any stage of plant development, neither as a support nor as a source of nutrients, because the water contains dissolved nutrients from which the plant will take them directly, achieving greater efficiency in the regulation of nutrition and fertilizer use [62]. The hydroponics that was worked was a drop system in this method tubes are used, emitters that drip the nutrient solution directly onto the roots of the plants with the growth, development and fruiting. It is controlled by a timer. This type of technique is ideal for the cultivation of plants with a specific nutrient demand, such as Solanaceae, Curbitaceae, *Capsicum* and *Lactuca*. The substrate, which was used in this work was hydroponics using a drop system, so the biocarbon and tepetzil substrates can be used.

The nutrient solution to be used is the Steiner solution (**Table 2**) [62], which, before its preparation, the water must be adjusted to a pH of 6.0 using sulfuric acid and phosphoric acid, taking phosphoric acid as a source of phosphorus in the solution. The nutrient solution was applied at 50% of its concentration during the first 45 days of plant growth, and then the Steiner solution was supplied at 100% at the beginning of flowering onwards.

A drip hydroponic system was used to supply the nutrient solution (**Table 3**). There will be eight irrigations during the day, with an interval of 1 hour between each one, starting at 08:00 a.m. and the last one at 03:00 p.m. Each irrigation will last 1 minute for the first 30 days and then increase to 2 min each irrigation from flowering onwards. Per irrigation, 280 mL/min of the solution will be supplied on average.

Regarding micronutrients, the commercial product Haifa Micro Combi was used, which contains Fe 7.4%, Mn 3.7%, Zn 1.1%, Cu 0.8% and Mo 0.5%; Nutri-Boro 11%, was used as a source of Boron, both products were added together with the Steiner solution through the hydroponic system (**Table 3**).

Nutrient solution (meq/L)					
Anions			**Cations**		
NO_3-	H_2PO_4-	SO_4^{2-}	K^+	Ca^{2+}	Mg^{2+}
12	1	7	7	9	4

Table 2.
Anions and cations for the nutrient solution [62].

Fertilizer	Quantities	
	meq/L	**g/1000 L**
Calcium nitrate	9	1060
Magnesium sulfate	4	494
Potassium sulfate	4	360
Potassium nitrate	3	303
Phosphoric acid	1	23 mL

Table 3.
Fertilizers for the nutrient solution.

4.1 Experimental

For the experiment, a completely randomized design was used with five treatments and 10 replications for a total of 50 experimental units, where three of the five treatments were the volumes of biochar mixed with tepetzil, in addition to treatment of only biochar and the control treatment which consists of only tepetzil (**Table 4**).

4.2 Fruit yield

The results were subjected to an analysis of variance (ANOVA) and a comparison of Tukey means ($P \leq 0.05$) using the statistical program Statistical Analysis System [63]. The total yield was obtained by adding the total weight of all the cuts per plant and treatment to subsequently extrapolate the yields per hectare (T ha^{-1}).

Regarding the yield with the hydroponic system, from which a total of 15 cuttings were obtained, a significant difference was found between the treatments, as shown in **Table 5**; the T3 (tepetzil 50% and biochar 50%) and T4 (tepetzil 25% and biochar 75%) treatments did not present significant differences between them, in turn, they were the treatments with the highest statistical yields with respect to the control treatment (T1), with a difference of 56.72 and 54.72 T ha^{-1} (T3 and T4, respectively) over the T1 treatment. The T2 (tepetzil 75% and biochar 25%) and T5 (biochar 100%) treatments are not significantly different from the T3 and T4 treatments, but neither from the control treatments. The yield of the *C. chinense* pepper grown in an NFT

Treatments	Tepetzil	Biochar
	%	
T1 (control)	100	0
T2	75	25
T3	50	50
T4	25	75
T5	0	100

Table 4.
Tepetzil and biochar substrate treatments and percentages.

Treatments	Tepetzil (%)	Biochar (%)	Yield (T ha^{-1})
1 (control)	100	0	37.95[b,‡]
2	75	25	44.35[ab]
3	50	50	56.72[a]
4	25	75	54.72[a]
5	0	100	49.64[ab]

[‡]*Means with the same literal are statistically equal according to the Tukey test (P ≤ 0.05).*

Table 5.
The yield fruit habanero in response to substrate mixtures of tepetzil and biochar.

(Nutrient Film Technique) hydroponic system without salinity restriction was 2.87 kg per plant, with an estimated yield of 46.1 Mg ha^{-1}, considering the area per plant adopted of 0.6225 m^2 [64].

5. Conclusions

The results show that the hydroponic system using the drip technique with the substrate mixed with 50% biochar and tepetzil showed a higher yield, obtaining 56.30 T ha^{-1}. Therefore, tropical vegetables often play a very important role in the Mexican economy. Therefore, the implementation of hydroponics as an innovative alternative form of crop production helps promote environmental protection as well as sustainability. Hydroponics, if this technology is applied to the production of tropical vegetables, does not require soil directly, but a substrate, in this case, biochar and tepetzil with the hydroponic system, mineral nutrients dissolved in water are provided, which is called nutrient solution, and the nutritional needs of habanero chili crops are covered.

Acknowledgements

The author would like to thank the Academic Division of Agricultural Sciences of the Juarez Autonomous University of Tabasco and the Academic Body of Sustainable Tropical Agriculture.

Conflict of interest

The authors declare no conflict of interest.

Notes/thanks/other declarations

Thanks, and no declarations.

Author details

Edmundo Gomez-Mendez, Pablo García-Lopez, Hortensia Brito-Vega* and Armando Gomez-Vazquez
División Académica de Ciencias Agropecuarias, Universidad Autónoma de Tabasco, Centro Villahermosa, Mexico

*Address all correspondence to: hortensia.brito@ujat.mx

References

[1] Tucuch-Haas CJ, Alcantar-Gonzalez G, Ordaz-Chaparro VM, Santizo-Rincon JA, Larque-Saavedra A. Production and quality of habanero peppers (*Capsicum chinense* Jacq.) with different NH_4 +/ NO^3 + substrate particle size ratios. Terra Latinoamericana. 2012;**30**(1):9-15

[2] Ramirez M, Arcos G, Mendez R, Meneses I. Chilli bell pepper varieties and hybrids for the tropics of Mexico. In: Meza VM, Chay AJ, Ramírez AR, Palacios RE, Valenzuela N, Alcantara JP, Kido MT, editors. Production Agriculture: An Approach Integrated. Mexico: University of Papaloapan; 2019. pp. 29-37. Available from: https://www.researchgate.net/publication/358749135_Produccion_Agropecuaria_Un_enfoque_integrado

[3] Castellanos JZ. Tomato Production Manual in Greenhouse. 1st ed. Intagri; 2009. pp. 456. Available from: https://www.intagri.com/public_files/Manual-de-produccion-de-tomate-bajo-invernadero.pdf

[4] Villa M, Catalan EA, Inzunza MA, Roman A, Macias H, Cabrera D. Hydroponic Production of Habanero Chilli in Greenhouses. Mexico: INIFAP; 2014. Available from: http://cenid-raspa.inifap.gob.mx/demo/modulo/Folletos%20tecnicos/2014/

[5] Hernandez-Hernandez M, Palma-Lopez DJ, Salgado-Garcia S, Palma-Cancino DJ, Rincon-Ramirez JA, Hidalgo-Moreno CI, et al. Charcoal as an improver of an acrisol cultivated with sugar cane (*Saccharum* spp.). Agro Productivity. 2020;**13**(5):35-41. DOI: 10.32854/agrop.vi.1607

[6] Ramirez FL. Forest charcoal as substrate, plus nitrogen, phosphorus and potassium (NPK), in the cultivation of *Raphanus sativa* radish [thesis]. Iquitos: National University of the Peruvian Amazon; 2014. Available from: https://repositorio.unapiquitos.edu.pe/handle/20.500.12737/3359

[7] Man JK, Yunho H, Juyeon L, Baknoon H, Arifur R, Hossain A, et al. Waste nutrient solutions from full-scale open hydroponic cultivation: Dynamics of effluent quality and removal of nitrogen and phosphorus using a pilot -scale sequencing batch reactor. Journal of Environmental Management. 2021;**281**:111893. DOI: 10.1016/j.jenvman.2020.111893

[8] Al SA. Hydroponics, aeroponic and aquaponic as compared with conventional farming. American Scientific Research Journal for Engineering, Technology, and Sciences. 2017;**27**(1):247-255. Available from: https://www.researchgate.net/profile/Ali-Shrouf-2/publication/318307802_Hydroponics_Aeroponic_and_Aquaponic_as_Compared_with_Conventional_Farming/links/658a78403c472d2e8e8d1d35/Hydroponics-Aeroponic-and-Aquaponic-as-Compared-with-Conventional-Farming.pdf

[9] Nemati MR, Simard F, Fortin JP, Beaudoin J. Potential use of biochar in growing media. Vadose Zone Journal. 2015;**14**(6):1-8. DOI: 10.2136/vzj2014.06.0074

[10] Huang L, Gu M. Effects of biochar on container substrate properties and growth of plants-a review. Horticulture. 2019;**5**(1):14

[11] Poma WA. Bibliographic review of the use of hydroponic systems in the cultivation of vegetables [thesis]. Peru:

César Vallejo University; 2020. Available from: https://repositorio.ucv.edu.pe/handle/20.500.12692/54960

[12] Perez-Cabrera CA, Juarez-Lopez P, Anzaldo-Hernandez J, Alia-Tejacal I, Gayosso-Rodriguez S, Salcedo-Perez E, et al. Rice husk biochar as a substrate for cucumber seedling growth. Journal Chapingo Horticulture Series. 2021;**27**(3):171-183. DOI: 10.5154/r.rchsh.2021.01.002

[13] Vargas-Tapia P, Castellanos-Ramos JZ, Sánchez-García P, Tijerina-Chávez L, López-Romero RMa, Ojodeagua-Arredondo JL. Physical, chemical and biological characterization of coconut dust substrates. Journal Phytotecnia Mexicana. 2008;**31**(4):375-381. DOI: 10.35196/rfm.2008.4.375

[14] Bu X, Ji H, Ma W, Mu C, Xian T, Zhou Z, et al. Effects of biochar as a peat-based substrate component on morphological, photosynthetic and biochemical characteristics of rhododendron delavayi Franch. Scientia Horticulturae. 2022;**302**:111148. DOI: 10.1016/j.scienta.2022.111148

[15] Guo Y, Niu G, Starman T, Volder A, Gu M. Poinsettia growth and development response to container root substrate with biochar. Horticulture. 2014;**8**(1):14. DOI: 10.3390/hortícolae4010001

[16] Awad YM, Lee SE, Ahmed MBM, Vu NT, Farooq M, Kim IS, et al. Biochar, a potential hydroponic growth substrate, enhances the nutritional status and growth of leafy vegetables. Journal of Cleaner Production. 2017;**156**:581-588. DOI: 10.1016/j.jclepro.2017.04.070

[17] Blok C, Van der Salm C, Hofland-Zijlstra J, Streminska M, Eveleens B, Regelink I, et al. Biochar for horticultural rooting media improvement: Evaluation of biochar from gasification and slow pyrolysis. Agronomy. 2017;**7**(1):6. DOI: 10.3390/agronomy7010006

[18] Kaudal BB, Chen D, Madhavan DB, Downie A, Weatherley A. An examination of physical and chemical properties of urban biochar for use as growing media substrate. Biomass and Bioenergy. 2016;**84**:49-58. DOI: 10.1016/j.biombioe.2015.11.012

[19] Lopez R, Mirafuentes HF, Aviles WI, Jimenez JA, Meneses I. Production of Habanero Chilli with Irrigation by Drip. Mexico: INIFAP; 2015

[20] Servicio Nacional de Inspección y Certificación de Semillas. Chile (Capsicum spp.). National Seed Inspection and Certification Service. Gobierno de México; 2017. Available from: https://www.gob.mx/snics/acciones-y-programas/chile-capsicum-spp

[21] Muñoz LS. Selection of varieties of habanero chilli (*Capsicum chinense* Jacq.) for industrial purposes [thesis]. Yucatan: Centro de Investigación Cientifica de Yucatan A.C; 2020. pp. 146. https://cicy.repositorioinstitucional.mx/jspui/bitstream/1003/1827/1/PCB_D_Tesis_2020_Liliana_Sarai_Mu%C3%B1oz_Ram%C3%ADrez.pdf

[22] Shared RT. Havana Chile, with Designation of Origin. 2017. Available from: http://www.gob.mx/firco/articulos/chile-habanero-con-denominacion-de-origen?idiom=es

[23] Hernandez J. It Is Consolidated he Habanero Chilli Looking for Exceed a Thousand Hectares of Sowing. SIT/Diario Presente; 2018. Available from: https://www.diariopresente.mx/tabasco/se-consolida-el-chile-habanero-buscan-superar-mil-hectareas-de-siembra/207960

[24] Marcial L. Growth and production of habanero chilli (*Capsicum chinense* Jacq.) under three sowing dates in the Comarca Lagunera [thesis]. Mexico: Autonomous University Antonio Narro Agrarian; 2016

[25] Lopez-Lopez R, Inzunza-Ibarra MA, Fierro-Alvarez A, Palma-Lopez DJ. Transplant dates and productivity of habanero chilli under irrigation by drip. Mexican Journal of Sciences Agricole. 2018;**9**(1):51-64. DOI: 10.29312/remexca.v9i1.847

[26] Fideicomiso de Riesgo Compartido. Chile Habanero, con Denominación de Origen. Gobierno de México; 2016. Available from: https://www.gob.mx/firco/articulos/chile-habanero-con-denominacion-de-origen?idiom=es

[27] National Research Institute Forestry, Agriculture and Livestock. Products and Services. Chile. 2017. National Research Institute. Forestry, Agriculture and Livestock. International Biochar Initiative. FAQs; 2013. Available from: https://biochar-international.org/resources/ibi-publications/

[28] Ramirez GS. Analysis of the production of habanero chilli (*Capsicum chinense* Jacq.) in the main states producers in Mexico during 2006-2015 [thesis]. Mexico: Autonomous University Agrarian Antonio Narro; 2017. Available from: http://repositorio.uaaan.mx:8080/xmlui/handle/123456789/42753

[29] Lopez M, Poot JE, Cortez MA. Response of habanero chilli (*Capsicum chinense* L. Jacq) to fertilizer supply organic in Tabasco, Mexico. Magazine UDO Agricole Scientist. 2012;**12**(2):307-312 http://repositorio.uaaan.mx:8080/xmlui/handle/123456789/42383

[30] Guillen RI. Fabricación de Carbón Vegetal Aprovechando los Raleos en Bosques Energéticos. Project Profile. Sistema De La Integracion Centroamericana. Alianza Energía Ambiente. San Salvador, El Salvador; 2008. p. 30. Available from: https://www.sica.int/busqueda/busqueda_archivo.aspx?Archivo=dpro_86588_1_30052014.pdf

[31] Diaz M, Gonzales A, Sifuentes D, Gonzales E. The charcoal: Alternative energy and chemical products. Xylem. 2010;**23**(1):95-103

[32] Cox J, Downie A, Hickey M, Jenkins A, Lines-Kelly R, McClintock A, et al. Biochar in Horticulture: Prospects for the Use of Biochar in Australian Horticulture. NSW Trade and Investment. Departments of Primaty Industries; 2012. p. 104. Available from: https://www.dpi.nsw.gov.au/__data/assets/pdf_file/0008/447857/DPI-BioChar-in-Horticulture.pdf

[33] Trelles WL, Gonzalez JL. Response of strawberry (*Fragaria vesca*) crop to different charcoal granulometries and soil moisture tension ranges under greenhouse conditions, at la Argelia experimental station [thesis]. Ecuador: National University of Loja; 2014. Available from: https://dspace.unl.edu.ec//handle/123456789/11795

[34] Valverde JC, Arias D, Campos R, Guevara M. Characterization physical and chemical characterization of charcoal from three stem segments and branches of *Eucalyptus camadulensis* from dendroenergy plantations. Mesoamerican Forestry Journal Kurú. 2018;**15**:16-24. DOI: 10.18845/rfmk.v15i1.3774

[35] Fiallos-Ortega LR, Flores-Mancheno LG, Duchi-Duchi N, Flores-Mancheno CI, Baño-Ayala D, Estrada-OrozcoL.Ecologicalsoilrestoration using biochar (charcoal), and its effect on *Medicago sativa* production. Science

in Agriculture. 2015;**12**(2):13-20. DOI: 10.19053/01228420.4349

[36] CABI, Authors Info & Affiliations. *Capsicum annuum* (Bell Pepper). CABI Compendium; 2019. DOI: 10.1079/cabicompendium.15784

[37] Pinedo JJ. Charcoal application in four treatments for nutrient recovery in agricultural soils [thesis]. Peru: Cesar Vallejo University; 2017. Available from: https://repositorio.ucv.edu.pe/handle/20.500.12692/33170

[38] Rios CL. Application of *Mespilus germanica* L. and *Mangifera indica* L. biochar in contaminated soils to reduce lead in the area of La Huaca-Huaral. Peru: Cesar Vallejo University; 2018. Available from: https://repositorio.ucv.edu.pe/handle/20.500.12692/18276

[39] Sanchez-Reinoso AD, Avila-Pedraza EA, Restrepo-Diaz H. Use of biochar in agriculture. Acta Biologica Colombiana. 2020;**25**(2):327-338. DOI: 10.15446/abc.v25n2.79466

[40] Dispenza V, De Pasquale C, Fascella G, Mammano MM, Alonzo G. Use of biochar as peat substitute for growing substrates of euphorbia × lomi potted plants. Spanish Journal of Agricultural Research. 2016;**14**(4):11. DOI: 10.5424/sjar/2016144-9082

[41] Escalante A, Perez G, Hidalgo C, Lopez J, Campo J, Valtierra E, et al. Biochar I: Nature, history, manufacture and use in the soil. Terra Latinoamericana. 2016;**34**(3):367-382

[42] Schahczenski J. Biochar and Sustainable Agriculture. ATTRA Sustainable Agriculture; The National Center for Appropriate Technology. 2018. pp. 12. Available from: https://attra.ncat.org/publication/biochar-and-sustainable-agriculture/

[43] Lampel P. Biochar from Garden Waste as a Potential Hydroponic Growing Substrate. Digitala Vetenskapliga Arkivet; Linköping University, Department of Management and Engineering Master's thesis. 2022. pp. 65. Available from: https://www.diva-portal.org/smash/get/diva2:1683930/FULLTEXT01.pdf

[44] Sarauer JL, Coleman MD. Biochar as a growing media component for containerized production of douglas-fir. Canadian Journal of Forest Research. 2018;**48**(5):581-588. DOI: 10.1139/cjfr-2017-0415

[45] Diaz-Perez M, Camacho-Ferre F. Effect of composts in substrates on the growth of tomato transplants. HortTechnology. 2010;**20**(2):361-367. DOI: 10.21273/HORTTECH.20.2.361

[46] Zarate MA. Hydroponics Manual. 2014. Available from: https://www.gob.mx/cms/uploads/attachment/file/232367/Manual_de_hidpononia.pdf

[47] Rajaseger G, Chan LK, Tan YK, Ramasamy S, Khin CM, Amaladoss A, et al. Hydroponics: Current trends in sustainable crop production. Bioinformation. 2023;**19**(9):925-938. DOI: 10.6026/97320630019925

[48] Rajendran S, Domalachenpa T, Arora H, Li P, Sharma A, Rajauria G. Hydroponics: Exploring innovative sustainable technologies and applications across crop production, with emphasis on potato mini-tuber cultivation. Heliyon. 2024;**10**(5):e26823. DOI: 10.1016/j.heliyon.2024.e26823

[49] Eden Green Technology. Hydroponics vs. Traditional Farming: A Comprehensive Comparison. Ministry of Agriculture and Rural Development; 2019. Available from: https://www.edengreen.com/blog-collection/hydroponics-vs-traditional-farming

[50] AgriTech Tomorrow. Water Management in Hydroponics: Techniques, Challenges, Solutions. Indoor & Vertical Farming, Precision Farming Grow Director Ltd.; 2024. Available from: https://www.agritechtomorrow.com/news/2024/03/31/water-management-in-hydroponics-techniques-challenges-solutions/15419/

[51] GMC C. Determination and study of substrates and fertigation in the production of certified citrus plants [thesis]. Mexico: Colegio de Postgraduados; 2012. Available from: http://colposdigital.colpos.mx:8080/xmlui/handle/10521/1833

[52] Service Mexican Geological Survey. Tepojal Monograph. SGM. Mexico: Service Mexican Geological Survey; 2016

[53] Osuna Fde J, Garcia F, Ramirez S, Canul J, Moreno MF. Substrate Management for Biological Control of Root Rot in Indoor Poinsettia with *Trichoderma* Spp. Mexico: INIFAP; 2011. Available from: https://www.compucampo.com/tecnicos/manejosustratos-nochebuenainterior.pdf

[54] Jaenicke H. Section 3: Substrate. In Good Tree Nursery Practices: Practical Guidelines for Research Nurseries. World Agroforestry Centre; 1999. pp. 30-44. Available from: https://apps.worldagroforestry.org/Units/Library/Books/PDFs/17_Good_tree_nursery_practices.pdf?n=44

[55] Pastor JN. Use of substrates in nurseries. Terra Latinoamericana. 1999;**17**(3):231-235. Available from: https://www.redalyc.org/pdf/573/57317307.pdf

[56] Burés S, Gago MC, Morales O, Marfà O, Xavier MF. Water characterization in granular materials. Acta Horticulturae. 1997;**450**:389-396. DOI: 10.17660/ActaHortic.1997.450.48

[57] Gayosso-RodriguezS, Borges-GomezL, Villanueva-Couoh E, Estrada-Botello MA, Garruña-Hernandez R. Substrates for growing flowers. Agrociencia. 2016;**50**(5):617-631. Available from: https://www.scielo.org.mx/pdf/agro/v50n5/1405-3195-agro-50-05-617.pdf

[58] Perez-Cabrera CA, Juarez-Lopez P, Anzaldo-Hernandez J, Alia-Tejacal I, Salcedo-Pérez E, Balois-Morales R. Potential benefits of biochar on agricultural crop productivity. Mexican Journal of Agricultural Sciences Agricultural. 2021;**12**(4):713-725. DOI: 10.29312/remexca.v12i4.2542

[59] Wiggins Z, Akaeze O, Nandwani D, Witcher A. Substrate properties and fertilizer rates on yield responses of lettuce in a vertical growth system. Sustainability. 2020;**12**(16):6465. DOI: 10.3390/su12166465

[60] Schafer G, Lerner B. Physical and chemical characteristics and analysis of plant substrate. Ornamental Horticulture. 2022;**28**(2):181-192. DOI: 10.1590/2447-536x.v28i2.2496

[61] Mixquititla-Casbis G, Villegas-Torres OG, Andrade-Rodriguez M, Sotelo-Nava H. Physical and chemical properties of substrates as a function of their particle size and organic-mineral component. Agricultural and Livestock Act. 2022;**8**(1):e0081007. DOI: 10.30973/aap/2022.8.0081007

[62] Steiner AA. The universal nutrient solution, proceedings of IWOSC. In: 6th International. Wageningen, The Netherlands: Congress on Soilless Culture; 1984. pp. 633-650. Available from: https://www.scirp.org/reference/referencespapers?referenceid=1833796

[63] University of North Carolina. Statistical Analysis System (SAS for Windows). University of North Carolina. Institute Inc., SAS Institute Inc.; 2023. Available from: https://www.sas.com/es_mx/learn/academic-programs/software.html

[64] Amorim BMA, Miler ST, Watanabe CAM, da Silva PVP, Raj GH, Severo RMR, et al. Hydroponic production of ' Biquinho' pepper with brackish water. Agricultural Water Management. 2021;**245**:106607. DOI: 10.1016/j.agwat.2020.106607

Chapter 3

Perspective Chapter: Best and Sustainable Practices for a Modern Cultivation Technique

Guangjae Lee, Jeongwook Heo and Dongeok Kim

Abstract

Consumers prefer high-quality agricultural products due to increased interest in income, lifespan, and health. We should be interested in the stable production of agricultural products with high-quality functional materials. Various control facilities that combine hydroponics or plant factory technology with the Internet of Things (IoT), artificial intelligence (AI), and deep learning will be introduced. Due to labor shortages, automation element technologies for labor reduction in hydroponics, aquaponics, and plant factories are expected to advance. In order for agricultural products produced in plant factory systems to be competitive with similar agricultural products produced in the field, new cultivation technologies are needed to increase plant productivity and reduce production costs. There is a need to develop recycling technology for badges and nutrient solutions. Improving sales and profitability through improved light efficiency and shortened growing seasons are important for maintaining the sustainability and profitability of the plant factory industry.

Keywords: hydroponics, LED, plant factory, nutrient solution, smart

1. Introduction

Recently, rapid climate change, including high temperatures, droughts, heavy rains, or heavy snowfalls, has emerged as a serious problem in terms of agricultural production and food security. Hydroponics and facility cultivation are being focused on avoiding serious environmental risks and protecting crops for stable production. It is estimated that the initial facility cultivation in Korea was performed about 500 years ago.

The early stages of agricultural facilities have developed into hot caps, plastic mulching, plastic tunnels, plastic houses, glass houses, precision agriculture, plant factories, and smart farms.

In recent years, precision agriculture or controlled environment agriculture (CEA), has emerged as a technology-based crop production system consisting of various nutrient supply systems, including hydroponics, aeroponics, and aquaculture, which optimizes the use of resources such as water, energy, and space to produce crops throughout the year [1]. When growing plants in open fields, yields and quality vary depending on weather conditions and other external factors, but CEA is designed to avoid these effects [2].

Although food shortages exist worldwide, consumers in most countries prefer high-quality agricultural products due to agricultural technology development and rising incomes. Farmers also tend to produce high-quality agricultural products to meet consumer demand or increase their own income.

Facility cultivation and plant factories are isolated from the outside, so they can provide stable agricultural products without being affected by diseases, pest damage, climate, etc.

However, plant factory systems are difficult to commercialize because they require high installation and operation costs and advanced operating technology.

Among various photosynthetic factors, light plays the most important role not only in facility cultivation but also in plant factories. The light assimilation function provides the energy required for photosynthesis, which suggests that light is the ultimate energy source for green plant metabolism, and photosynthetic efficiency varies depending on the spectral wavelength (light quality) [3]. Light quality plays an important role in the transition of plants from vegetative growth to reproductive growth [4] and aging [5].

In this chapter, we will focus on hydroponics, plant factories, artificial lighting, and functional materials. The characteristics of LEDs as a light source and the characteristics of the light environment in plant factories with artificial lighting are discussed.

2. Product quality improvement

2.1 Specific ingredients as functional chemicals

2.1.1 Flavonoids

Flavonoids are secondary metabolites found in plant organs and tissues and are produced through complex interactions with plant survival and environmental factors. The main component that directly affects flavonoids is light as an energy source. Light controls plant growth stages and maintain photosynthetic efficiency through light intensity, quality, and irradiance [6].

Lee et al. [7] reported the total flavonoid content of lettuce was higher in the red 80/blue 40 μmol m^{-2} s^{-1} and R60/B60 μmol m^{-2} s^{-1} treatments (0.72 catechin mg g^{-1} FW). Flavonoids have a wide range of pharmacological activities [7]. The blue LED or a combination of red and blue LEDs increased the phenolic compound content of amaranth sprouts compared to fluorescent lighting [8]. In a closed plant factory, the total phenolic content of lettuce at 4 weeks was measured as 0.313, 0.262, and 0.383 mg GAE/plant in control, side anion treatment, and upwelling treatment groups, respectively [9]. When synthesizing various experimental results, total flavonoid content varies depending on the cultivation method and variety [7].

2.1.2 Phenolic compounds and antioxidants

It is generally known that there are about 4000 kinds of polyphenol substances in plants. Phenolic compounds are divided into three types: simple phenols, phenolic acids, hydroxycinnamic acid derivatives, and flavonoids, and these substances have a certain level of antioxidant, antimutagenic, and anticancer properties [10].

Lee et al. [7] reported the total phenol content differed depending on the cultivation method, with 117.84 mg·100 g^{-1} GE in aeroponics, 98.57 mg·100 g^{-1} GE in hydroponics, and 74.62 mg·100 g^{-1} GE in perlite cultivation. In addition, the highest phenol compound content was observed in the red-blue LED 80/40 μmol m^{-2} s^{-1} treatment (0.76 GAE mg·100 g^{-1} FW reached) [7]. The total phenol content of *Peucedanum japonicum* by light source was the highest at 105.77 mg·100 g^{-1} GE in RB, followed by 92.52 mg·100 g^{-1} GE in RBW, 89.08 mg·100 g^{-1} GE in fluorescent light (FL), and 82.00 mg·100 g^{-1} GE in RBGW, showing differences depending on the light source.

The extract of 3.08 mg g^{-1} in field cultivation of *Peucedanum japonicum* was higher than that of 2.39 mg g^{-1} in greenhouse cultivation [11]. The phenol content of red lettuce was highest when the LED R:B light intensity was 90:30 μmol m^{-2} s^{-1}, and that of green lettuce was highest when the LED R:B light intensity was 80:40 μmol m^{-2} s^{-1} [12].

In lettuce cultivation, when downward anion treatment was applied, total phenol content and productivity increased, but antioxidant activity decreased because the leaf area of lettuce was the largest [9]. Carrot leaves contain antioxidant components and are suitable as leaf vegetables [13].

2.1.3 Glucosinolate

Glucosinolate (GLS) is a representative phytoalexin that contains nitrogen and sulfur and is a secondary metabolite biosynthesized in approximately 100 species of Brassica plants [14, 15]. GLS is known to determine the unique taste and aroma of crops and exhibit various functional properties in vivo. Kale (*Brassica oleracea* L. *acephala*) was reported to be rich in glucosinolates (GSLs) [16], Kushad et al. [17] reported that among vegetables, kale (*Brassica oleracea* var. *acephala* DC.), Chinese kale (*B. oleracea* var. *alboglabra* Bailey), red cabbage (*B. oleracea* var. *capitata* f. *rubra* L.), Brussels sprouts (*B. oleracea* var. *germmifera* Zenker), kohlrabi (*B. oleracea* var. *gongylodes* L.), and broccoli (*B. oleracea* var. *italic* Plen) were rich in GSLs.

The main GLS in the sprouts 6 days after sowing of kale were progoitrin, gluconapine, sinigrin (SIN), glucoerucine, glucoraphamin (GLP), glucoiberin (GLB), and 4-hydroxyglucobrassicin [18], and it is thought that the difference in the contents of glucosinolate components in different growth stages of kale is due to the difference in contents in different growth stages [19]. When the LED light source was blue:white (1:1), red:blue:white (2:1:3), and blue:white + fluorescent light (FL) (1:1:1), the contents of GLS in all treatment groups in the plant factory kale cultivation were high in the order of glucosinolate, GLB, SIN, gluconasturtin, progoitrin, GLP, and epiprogoitrin regardless of the light source [19].

Photoperiod-dependent changes in phytochemicals of broccoli and GSL biosynthesis in *Arabidopsis thaliana* have also been reported [20, 21]. In kale shoot tissue, wavelengths of 640 nm and 730 nm were found to have high SIN content, but other wavelengths did not affect GSL accumulation [22].

Blue and white LEDs and FL showed high contents of PRO, EPI, SIN, GLP, and GLN, while red, blue, and white LEDs (RBW) showed the highest contents of GLB, GLBR, and total GLS [19]. The glucosinolate content of watercress was lowest in LED red:blue (9:1) and highest in LED red:blue (7:3). LED red:blue:green (1:1:1) had a negative effect on the growth of watercress, but played a positive role in increasing functional components [23].

2.1.4 Ascorbic acid

The ascorbic acid is a type of water-soluble vitamin that is sensitive to heat and is abundant in vegetables and fruits. Depending on the cultivation methods, the ascorbic acid content of *Peucedanum japonicum* was the highest in aeroponics at 108.23 $mg{\cdot}100\ g^{-1}$, followed by perlite at 88.05 $mg{\cdot}100\ g^{-1}$ and nutrient film technique (NFT) at 80.83 $mg{\cdot}100\ g^{-1}$, showing statistical significance depending on the treatment [24].

A high correlation between ascorbic acid and glucose content was reported in lettuce grown under 16 hours of light [25]. In addition, ascorbic acid in cabbage grown under blue LED increased [26]. Kim and Lee [27] reported that ascorbic acid in lettuce is greatly affected by photoperiod. Spinach ascorbic acid content increased with increasing cultivation days [28], but there was no significant change in spinach ascorbic acid content depending on harvest time [11].

The growth and ascorbic acid content of lettuce is significantly affected by supplemental UV-A LED irradiation [27]. The ascorbic acid content of cabbage increased with additional UV-A LED irradiation [29], and the ascorbic acid content of lettuce increased by 20–30% in blue and red LED+UV-A LED treatments compared to when no UV-A LED irradiation was performed [27].

It was shown that vitamin C content varies depending on the growth stage, harvest time, and cultivation method of the crop [11, 28]. This is thought to be because the vitamin C content of crops is a type of phytochemical and is a variable affected by the environment.

2.1.5 Amino acids

Cysteine content in *Peucedanum japonicum* was highest in aeroponics at 46.76 $mg{\cdot}100\ g^{-1}$, followed by NFT and perlite, and according to artificial lighting source, fluorescent lamps were the highest at 43.53 $mg{\cdot}100\ g^{-1}$, and LED red, blue, green, white types (such as RB, RBW, and RBGW) did not show statistical significance [24].

Methionine content according to *Peucedanum japonicum* cultivation method was highest in NFT and aeroponics, and lowest in perlite at 75.64 $mg{\cdot}100\ g^{-1}$, and according to artificial lighting source, methionine content was highest in FL at 79.43 $mg{\cdot}100\ g^{-1}$, and RB, RBW, and RBGW did not show statistical significance among treatment groups [24].

Kim and Choi [30], glutamic acid and aspartic acid in the elm tree were 58.5 $mg{\cdot}100\ g^{-1}$ and 53.8 $mg{\cdot}100\ g^{-1}$, respectively, accounting for 18.26% of the major amino acids. The methionine content was detected at 7.7 $mg{\cdot}100\ g^{-1}$, but only a trace amount of cysteine was detected. Depending on the type of crop, the amino acid content may vary depending on the cultivation method, even for the same crop [30].

2.1.6 Anthocyanin

Anthocyanins are a type of phytochemicals that constitute natural pigments in plants and are water-soluble substances of the flavonoid series that are abundantly contained in fruits, vegetables, and flowers. Among them, the light environment is known to be an effective environmental factor that promotes the production and expression of phytochemicals including anthocyanins [31, 32].

Anthocyanin synthesis varies depending on the type of crop as well as the light environment such as light quality, light intensity, and photoperiod. Cryptochrome, which acts as a photoreceptor for blue light and UV-A, is known to promote the expression of genes related to anthocyanin biosynthesis [32, 33].

As the photoperiod of plants shortens, anthocyanin accumulation increases in broccoli [21], but decreases in potatoes [34], and berries [35].

Anthocyanins accumulate relatively more under blue light conditions or shorter photoperiods. As the dark cycle decreased under blue LED, anthocyanin in lettuce significantly increased [27]. Lee and Kim [36] reported the accumulation of anthocyanin increased under red-blue LEDs of 634, 659, and 450 nm, and it was possible to produce red mustard with excellent growth and high anthocyanin content under light intensity of 300 μmol m^{-2} s^{-1} and photoperiod (h) of 16/8 using LED [37].

2.2 Nutrient elements in fruits

2.2.1 Calcium

Calcium is not easily transported within the plant, so calcium deficiency occurs frequently. It is known as the main cause of tip burn, which causes necrosis of young plant leaves and reduces marketability in plant factories [38]. Calcium is an important component of plant cell walls, and once fixed in the cell wall, it is immobile. Calcium cannot move between tissues with different calcium levels, and tissues often compete for the distribution of calcium delivered from the roots [39].

Calcium deficiency disorders are commonly observed in tissues with low transpiration rates, such as fruits and young leaves [40], and calcium transport is not related to the transpiration of fruits and the cation exchange capacity of fruit tissues. The apical transport of calcium absorbed from leaves is greater than basal calcium transport [41], and high relative humidity accelerates Ca transport in plants [42].

To increase the calcium content of hydroponically grown tomatoes, supplementation of Ca-gluconate ($C_{12}H_{22}CaO_{14}{\cdot}H_2O$, Ca-glu) was the most effective supplemental calcium source compared to calcium chloride and calcium carbonate [43], and the inclusion of 0.9 me L^{-1} Ca-glu in the nutrient solution could increase the calcium content of tomato fruits by 49% [42].

The Ca-glu leaf treatment could prevent calcium deficiency in green peppers and apples [44], and the fruits treated with higher amounts of calcium at the mature green stage of tomatoes had more cell wall components than fruits treated with lower amounts of calcium during ripening [45]. The common tomatoes lost 25% of cellulose and 53% of hemicellulose during ripening [45]. The content of alcohol-insoluble substances (AIS), a cell wall component, had a similar effect on tomatoes as calcium, and fruits with higher calcium content also had higher AIS contents [11]. The AIS composition of common tomatoes decreased, while that of fleshy tomatoes remained constant during ripening [46].

Tip burn requires skilled techniques, and early detection of tip burn can help to manipulate environmental parameters to improve the freshness of lettuce grown in plant factories [47]. Early detection of tip burn can help growers manipulate controlled environmental parameters to improve the freshness of lettuce grown in plant factories [47].

2.2.2 Selenium

Selenium has been applied to forage crops along with other fertilizers, along with some other crops in Finland and New Zealand [48]. Generally, low concentrations of sodium selenite are nontoxic and are easily absorbed by plant roots. Chen [49] reported that rice selenate fertilizer had 3.59% higher selenium content than rice treated with selenite fertilizer. In addition, when selenium was applied to the leaves, the yield, sweetness, flavor, total amino acid, and ascorbic acid content of green tea significantly increased [50].

The National Research Council of the United States announced the recommended daily intake of selenium as 50–200 μg [51], and sodium selenite (Na_2SeO_4) is added to the nutrient solution during hydroponic cultivation to increase the selenium content. The selenium content of Chinese leafy vegetables increased as the selenium concentration in the nutrient solution increased [52]. The selenium content and firmness of tomato fruits increased rapidly as the selenium concentration in the nutrient solution increased [53]. In addition, considering the fruit weight, available solids, and the National Research Council recommendation for the daily selenium intake of 50–200 μg for adults [51], the optimal selenium concentration in the tomato nutrient solution was 1 mg L^{-1} [53].

3. Hydroponics as labor savings

3.1 Hydroponics

Hydroponics is the cultivation of plants without soil and can be divided into cultivation using only water and cultivation using a medium. Hydroponics dates back to the eighteenth century and is thought to have existed before that [54]. The global hydroponics market forecast is estimated to rise at a CAGR of 13.7% over the forecast period. In 2023, the hydroponics industry share is expected to be valued at US$ 12,644 million. The market is expected to surpass US$ 45,623 million by 2033 [55].

In the early days of hydroponics, readily available and inexpensive substrates were used for hydroponics, but commercial substrates such as rockwool and cocopeat became mainstream. Hydroponics has the advantages of producing uniform and high-quality crops, increasing productivity due to faster growth, reducing labor, reducing water consumption, year-round cultivation, and reducing the area required for production, as crops are grown under optimal conditions [56].

Hydroponics has been used to improve the quality of high-quality lettuce [57], and tomatoes [58]. Hydroponic systems can predict yield and harvest time, improve fruit quality, and control nutrient supply and environmental conditions [59], but they require advanced technology and operational capabilities. The industry is growing in the United States as demand for fresh, locally grown, and safe food increases [55]. Hydroponic products have a better appearance, which attracts consumers. Precise control improves the shelf life of products under shelf and refrigerated conditions compared to conventional cultivation systems [60].

Various hydroponic systems such as nutrient film technology (NFT), drip system, aeroponics, Ebb Flow, hydroponics, and wick systems have gained popularity. Hydroponics and plant factories rely on machines and facilities based on sensors for most of the work previously done by humans. In particular, most environmental factors such as ventilation, temperature, shading, water, and fertilizer supply, can be controlled through computers, smartphones, and machines and devices.

Recently, a system that predicts disease occurrence through the leaves of crops has been developed and is in the practical stage. Plant factories are a form of advanced agricultural industry that has developed based on hydroponics technology.

About 10–15 years ago, there were reports that hydroponics or plant factories could reduce labor by about 30%, but now the technology has developed so quickly that it is meaningless to even mention it. Currently, crop cultivation is indirectly controlled through interfaces, but recently, various control facilities that combine the Internet of Things (IoT), artificial intelligence (AI), and deep learning will become mainstream.

Labor is reduced through optimal cultivation and optimal environments such as hydroponics, aquaponics, and plant factories. In the future, various technologies will be developed to reduce labor.

3.2 NFT system

Nutrient film technique (NFT) is a hydroponic technique where a very shallow stream of water containing all the dissolved nutrients required for plant growth is recirculated past the bare roots of plants in a watertight gully, also known as channels (https://en.wikipedia.org/wiki/Nutrient_film_technique) [Accessed: September 25, 2024].

NFT systems are either open systems that intermittently supply nutrient solutions when using a medium, such as for vegetables, or closed systems that circulate nutrient solutions when not using a medium, such as for leafy vegetables. NFTs require a high level of skill from the user, mainly in adjusting nutrient solution requirements according to the various growth stages of the plant.

NFT is an ideal system compared to other hydroponics or soil cultivation [61], and the optimal flow rate for improving the growth rate of lettuce hydroponics is 20 L/h [62]. Lettuce accounts for approximately 7% of the entire US industry [55], and lettuce is grown almost exclusively in hydroponics by NFT systems.

The water flow in NFT must be balanced with other factors such as temperature, pH, electrical conductivity (EC), slope of the path, and concentrations of nitrogen, potassium, and calcium to ensure that the plants can sufficiently absorb the nutrients they need for growth. NFT has also been used in lettuce, garlic, potatoes, tomatoes, cucumbers, strawberries, and beans [63]. Total yield and commercial yield of long-ripened tomatoes varied with cultivar and growing period ($P < 0.01$), indicating that cultivar performance was affected by growing period [64].

3.3 Aquaponics

Small-scale aquaponics systems feature aquariums and plant cultivation facilities and are also used indoors to create esthetically pleasing indoor spaces. Aquaponics is a technology that combines aquaculture and hydroponics into one system, using nutrient-rich water from aquaculture to grow plants [65]. In aquaponics systems, most nutrients essential for plant growth can be supplied from fish feed. The pH of the system plays an important role in nutrient concentration, as well as influencing the accumulation of various nutrients. This is a conventional system that takes into account water volume and nutrient utilization [66].

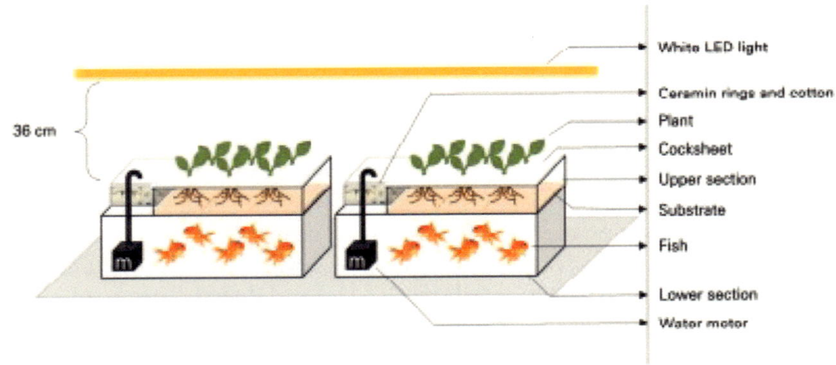

Figure 1.
Schematic diagram of the aquaponics tank where plants and fish grow together [69].

The water quality of an aquaponics system is very important because it is a limited factor. It is also necessary to maintain adequate water quality to provide optimal conditions for plant growth and a healthy environment for fish to ensure a successful system. Nevertheless, plants and fish have different water quality requirements. In particular, EC strength and water pH are key factors for potential aquaponics systems. The pH of hydroponic water plays an important role in fish health through the oxidation of NO_2 and NH_4 in this system [67]. Schreier et al. [68] mentioned that fish waste and fish feed in hydroponic systems can release H^+ through nitrification by adding ammonia to the hydroponic effluent.

Geranium (*Pelargonium zonale*) and goldfish (*Carassius auratus auratus*) were fed 0.4 g of fish feed per day (**Figure 1**). The diet components were calcium (0.40%), phosphorus (1.80%), crude fat (4.0%), crude fiber (5.0%), crude meal (13.0%), and crude protein (36.0%). $KHCO_3$ + Fe-EDTA treatment was more desirable for sustainable water quality and the growth of geranium plants in hydroponic systems [69]. Choi et al. [69] suggested that future studies should investigate the effects of fish and feed densities on the hydroponic growth of aquatic plants, plant bioactive compounds, and plant morphological growth in response to fish.

4. Plant factory

4.1 Plant factory

Plant factories are systems that combine hydroponics and IT technologies to artificially control and regulate environmental factors such as temperature, humidity, light, and carbon dioxide to produce crops all year round. Recently, climate change has become a problem that threatens the stable production of crops, and interest in alternative cultivation methods is growing [7, 70]. In particular, plant factories have gained popularity as a solution to three similar undesirable alternatives: food shortage or unstable supply, resource shortage, and environmental destruction [71].

However, there are also disadvantages, such as the initial installation cost of the plant factory system, the need for skilled labor, and the rapid spread of diseases when pests are introduced into the plant factory system [54]. Fully controlled light-growth plant factories require a lot of electrical energy for artificial lighting for crop cultivation, while they have high yields and reduced field management labor [72].

Ikeda et al. [73] reported that lettuce grown under optimal conditions grows eight times faster than lettuce grown under normal conditions, and this rapid growth rate is the main cause of physiological disorders related to mineral nutrition.

By cultivating medicinal crops with high demand through customized optimal environmental management for each crop, high-quality, high yields, and year-round cultivation through shortened cultivation periods are possible [74], so it is worth actively attempting to cultivate medicinal plants with high added value in plant factories.

Since the amount of physiologically active substances in plants varies depending on the cultivation method, it is necessary to study the optimal cultivation conditions for mass production by varying the harvest period and cultivation conditions [75].

Improving sales and profitability through improved light source efficiency and shortened growing periods is important for creating sustainability and profitability in the plant factory industry, and ragweed was selected as a recommended crop in the Controlled Ecological Life Support System project promoted in the United States [76].

New cultivation technologies are needed to increase plant productivity and reduce production costs so that agricultural products produced in plant factory systems can compete with general agricultural products produced in the field. Recently, various leafy vegetables have been produced in plant factories, but as an alternative to the initial installation and operating costs of plant factories, it is important to cultivate value-added crops that save electric energy and increase income [77], and to discover and cultivate functional materials and pharmaceutical materials using LED lights of various wavelengths.

4.2 Control of light conditions

Environmental factors such as light, temperature, humidity, CO_2 concentration, and nutrient supply have a great influence on the growth and development of plants. Among them, light is the main energy source for photosynthesis and is considered a major factor that regulates the growth and development of plants as a stimulus signal [78].

Sunlight has various wavelengths of light and is free, but its quality and photoperiod are difficult to control. On the other hand, light conditions such as photoperiod can be easily controlled using artificial lighting in plant factories [79].

When artificial light sources such as trichromatic lamps, sodium lamps, and red LEDs were used to supplement light before sunrise and after sunset, red LEDs supplemented with 1.2 $\mu mol\ m^{-2}\ s^{-1}$ reduced electrical energy consumption and significantly increased the yield of kale leaves [79].

Fully controlled photo-growing plant factories enable high-quality, high-yield cultivation all year round but artificial lighting for crop cultivation requires a lot of electrical energy [72]. Therefore, it is important to increase light use efficiency (LUE) in plant factories, and a multifaceted approach is required. To improve LUE, light quality was changed [80, 81], and crop productivity was improved by controlling photoperiod [82].

The quality of light irradiated to plants is known to affect plant growth and development through selective activation of various photoreceptors [83]. Light causes differences in growth and secondary metabolite synthesis [19], with red and blue wavelengths being effectively absorbed by chlorophyll and used for photosynthesis [84].

4.2.1 LED

LEDs can produce special wavelengths, which can have various effects on plants. Although LEDs are expensive, they have many advantages, such as excellent performance, a relatively high energy conversion factor, a wide spectrum, a relatively low surface temperature, and a long lifespan [85]. Improved lighting efficiency and shortened crop-growing seasons are essential to maintaining the sustainability and profitability of the plant factory industry.

Fully controlled photo-growth plant factories enable high-quality, high-yield crops all year round but require a lot of electrical energy for artificial lighting for crop cultivation [72]. Therefore, increasing the light use efficiency (LUE) in plant factories is important and requires a multifaceted approach. Light quality has been changed to improve LUE [80, 81, 86], and crop productivity has been improved by controlling photoperiod [82].

The cumulative electrical energy consumption is reduced by 75.6% for red LEDs and 70.7% for blue LEDs compared to fluorescent lamps [87]. Dorais et al. [88] reported that tomato yields increased at high light intensities (150 μmol m^{-2} s^{-1}). Improving light efficiency and shortening crop growth periods to improve sales and profitability are most important for maintaining the sustainability and profitability of the plant factory industry.

LEDs can selectively irradiate wavelengths suitable for crop photosynthesis and are capable of close-range lighting, and the peak wavelengths of LEDs are red light in the range of 630–670 nm and blue light in the range of 430–470 nm [89–91]. The effects on plant growth responses and secondary metabolite yields vary depending on the light source.

Red LEDs promote plant biomass increases such as fresh weight, dry weight, plant height, and leaf area [90], and affect the growth of the aboveground part of lettuce [92].

Blue LEDs have been found to induce chlorophyll formation and development without directly affecting plant growth, and blue light has been associated with underground growth in lettuce [92].

Green light has a higher reflectance when irradiated to plants than other wavelengths and has been considered to be less useful for plant growth, especially photomorphogenesis and photosynthesis [92].

Red and blue wavelengths of light are effectively absorbed by chlorophyll and used for photosynthesis [84]. UV rays that cause stress to plants are divided into three regions according to wavelength: UV-A (315–380 nm), UV-B (280–315 nm), and UV-C (100–280 nm). UV-A and some UV-B contained in solar radiation reach the Earth's surface, but UV-C is blocked by the ozone layer and does not reach the Earth's surface. Therefore, UV-A accounts for approximately 95% of the UV rays reaching the Earth's surface. On the other hand, in greenhouses, the intensity of the transmitted UV rays is drastically reduced due to the characteristics of the covering material that absorbs UV rays [93].

4.3 Pulse, duty rate, and photoperiod

The effects on plant growth responses and secondary metabolite yields vary depending on the light source. Cho et al. [94] also suggested that the Photosynthetic Photon Flux Density (PPFD) level (≥100 μmol m^{-2} s^{-1}) of pulsed LEDs was an important factor affecting plant growth and LUE. Despite the decrease in PPFD, none of the pulsed LEDs significantly reduced the SPAD value. Pulsed LEDs with millisecond cycles did not inhibit chlorophyll biosynthesis in wheat leaves during vegetative growth [95].

Most of the early studies compared the efficiency of pulsed LEDs and continuous LEDs (control) at different duty ratios at the same light intensity. In plant factories, LEDs can be controlled by intermittent and pulsed lighting [96]. The ultimate goal of using pulsed LEDs in horticulture is not simply to explore the effects of pulsed irradiation on crop growth and development but to save energy without growth inhibition.

The light use efficiency was highest when irradiated with 1 kHz pulsed LEDs [97]. Simultaneous pulse irradiation of red and blue LEDs is known to be more effective in improving growth and development than alternative pulse irradiation using red and blue LEDs. Pulsed LED light quality also affects plant growth and development, and pulsed red light increases the length and fresh weight of hypocotyls, and pulsed blue light induces chlorophyll a, b, and carotenoid accumulation [98]. LEDs can be controlled in plant factories with intermittent and pulsed lighting [96].

Son et al. [97] reported that pulsed LEDs with a duty ratio of 75% and a low frequency did not significantly inhibit plant growth, suggesting that pulsed LED irradiation technology has the potential to reduce energy consumption during crop production in plant factories.

Photoperiod affects flowering responses, growth, and development of plants [99], and the growth of leafy kale seedlings was significantly affected by supplemental radiation during shortened daylight hours in a greenhouse [100].

4.4 Electrical conductivity (EC)

Electrical conductivity (EC) is used as an auxiliary means of measuring the concentration of fertilizer salts in a nutrient solution and is the amount of ions that conduct electricity in the culture medium. EC measurement is an indirect and rapid method to measure the total concentration of ions, including dissolved nutrients, in a solution [101, 102]. Electrical conductivity does not mean that the fertilizer components are present in the same proportions when the nutrient solution is first prepared. Since plants have different absorption tendencies depending on the type of fertilizer component and pH, electrical conductivity and pH should be measured regularly.

Generally, leafy vegetables have a rating of 1.0–1.5 dS m^{-1}, fruits and vegetables have a rating of 2.0–3.0 dS m^{-1}, and for special purposes, the rating is temporarily increased to 4.5 dS m^{-1} for cultivating fruit crops. Typically, EC is maintained in the range of 1.5–3.5 dS m^{-1} to obtain optimal yields [103]. If the EC range of the crop is higher than the ideal range for hydroponic cultivation, nutrient absorption is hindered due to increased osmotic pressure, while low EC can seriously affect plant health and yield.

Samarakoon et al. [104] reported that the absorption rate of all nutrients increased with increasing nutrient solution concentration (EC), but did not contribute to the increase in yield, but the optimal EC for 20–25 days of hydroponic cultivation for carrot leaf production was reported to be 3.0 dS m^{-1} [13]. Supplying 1.5 dS m^{-1} of EC during the seedling and flowering stages of tomatoes and 4.5 dS m^{-1} of EC during the fruiting and harvest stages significantly promoted growth and improved yield and quality [105]. High uptake of ions may have caused nutrient imbalance in plants, which may have reduced total dry matter production. EC above a certain level increased N, P, K, and Ca uptake by plants but did not significantly increase leaf growth and yield [104].

4.5 Plant density

The planting density varies depending on the characteristics of the crop, and most leafy vegetables are planted at 15–20 × 15–25 cm spacing to secure light for photosynthesis and space for the leaves to grow to an appropriate size. The linear relationship between the dry weight of new shoots and the fresh weight was observed to be linear regardless of the planting distance. In the explanatory function equation, the crop growth rate, relative growth rate, and loss time showed a quadratic function form [106].

4.6 Automation

4.6.1 Plant automation

Plant automation refers to the use of technology and control systems to automate and optimize the processes and equipment in a manufacturing or industrial plant. Plant biotechnology can be defined as the use of tissue culture and genetic engineering techniques to produce genetically modified plants that exhibit new or improved desirable characteristics (https://www.plantengineering.com/plant-automation/) [Accessed: September 25, 2024].

4.6.2 Benefit of automation of plant operations

Increased productivity and efficiency: Automated systems can work faster and more accurately than manual labor, leading to increased output and reduced waste.

Improved quality control: Automated systems can be programmed to perform specific tasks with high precision and consistency, which can lead to improved product quality.

Reduced labor costs: Automation can reduce the need for human labor, which can lower labor costs and increase profitability.

Enhanced safety: Automated systems can reduce the risk of accidents and injuries in the workplace.

Increased flexibility: Automated systems can be easily reprogrammed to adapt to changing production needs, which can increase the flexibility of plant operations.

Remote monitoring and control: Automated systems can be controlled and monitored remotely, which can reduce downtime and improve response times to problems (https://www.plantengineering.com/plant-automation/) [Accessed: September 25, 2024].

4.7 Plant factory modules

Plant factory automatic system consists of a drive module that transports the tray, an environmental sensing module that controls the environment, a vision module that monitors the growth, and a main computer module.

The drive module controls the precise location of the tray movement, such as the entrance of the tray which planted plants, selecting and storing to grow, and exit for harvesting. The environmental sensing module consists of a number of temperature sensors, humidity sensors, optical sensors, and CO_2 sensors in the cultivation room. It monitors and controls the environment.

The vision module is a monitoring system installed in the robot's shuttle and monitors the degree of growth from up and down, front and rear, right and left, and all directions through the office.

The main computer module is a main control system that collects information from the drive module, environmental module, and vision module. It takes charge of all control of the room according to work instructions and processes data from the collected information (https://www.famecs.com/eng/view.do?no=414) [Accessed: September 25, 2024].

5. Reduced environmental load and new crops

5.1 Recycling nutrient solution

Nutrient solutions are commonly recycled during hydroponic cultivation to reduce waste [107, 108]. Nutrient imbalances can result in an oversupply of ions such as SO^{-4}, Ca^{+2}, and Mg^{+2}, which can affect growth and yield and increase electrical conductivity (EC), necessitating disposal of the solution [109].

Quality irrigation water can be a limiting factor in many commercial hydroponic greenhouses [110]. Some large commercial operations routinely analyze recycled solutions for individual nutrient levels in the laboratory, but most farmers prefer to discard recycled solutions [104, 111].

However, nutrient solution recycling can minimize fertilizer waste and reduce environmental pollution caused by frequent disposal. Image analysis of the canopy area of the plant community showed that the negative effects of recycling began 2 weeks after transplanting and were not observed when the recycled solution was discarded after 2 weeks or when reverse osmosis water was continuously used [112].

Considering the ion balance of the recycled nutrient solution in hydroponic cultivation of paprika, it was most effective to replace it every 4 weeks, and considering the yield and nutrient and water absorption, it was also possible to replace it every 8 weeks [113].

Even when the nutrient solution was recycled, the ion accumulation and salinity increased dramatically, which had a short-term negative effect on fruit yield and fruit quality [113]. Therefore, it is necessary to replace the nutrient solution periodically during the growing period.

Nutrient solution recycling can often improve the efficiency of mineral nutrient solution use and reduce soil and groundwater contamination outside the greenhouse. However, recycling can increase the density of specific microorganisms in the nutrient solution or cause nutrient imbalance due to the accumulation of specific ions, which affects crop yield and quality [113].

The EC of the recycled solution is measured to determine nutrient status [114]. Maintaining the EC of the recycle solution at a target level does not necessarily result in optimal concentrations of individual plant nutrients [102]. A recycling system based on EC control was tested and compared to a system without recycling, and it was found that rose flower yield and quality were not affected by recycling [115]. There was a good relationship between the EC and ion concentrations of the supplied solution, the recycled leachate solution, and the non-recycle leachate solution.

5.2 Recycling medium

Among artificial substrates, rockwool has excellent physical and chemical properties and is widely used in crop production [116]. Rockwool is expensive to purchase and difficult to dispose of after use. To compensate for these shortcomings, studies have been conducted to reuse the substrate [117].

Cocopeat is a popular substitute for rockwool, which is made from the outer shell of coconut and can be used as compost after use [118]. When the coir substrate was recycled as a seedling substrate after 2 years of hydroponic tomato cultivation, the growth was best when the coir content of the substrate was 75% or more for cabbage and 85% or more for lettuce [119]. Hydroponic cultivation accounts for more than 75% of the organic substrate cultivation area in China, and it can save water and fertilizer, increase yield, and protect the environment [120, 121].

In closed and open hydroponic systems, the physicochemical properties of the root medium have a significant impact on strawberry growth and yield [122], and the root medium changes its physicochemical properties due to the continuous decomposition of the substrate components. These particles can be decomposed by physical forces or microbial influences [123].

Due to the limited root volume and low ion buffering capacity, irrigation water and nutrient content must be carefully controlled [103, 118]. Many formulations of essential macro- and micronutrients have been developed to enhance plant growth and nutrient uptake [124, 125]. Controlling the chemical composition and concentration of nutrient solutions is a key technology for recycling substrate cultivation.

5.3 Management drainage solution

The changes in the chemical characteristics of the drainage solution as well as the growth, yield, and fruit quality were investigated in relation to the number of years the root media used. The changes in nutrient composition of the drainage solution and their effect on tissue nutrient content can be used to modify the nutrient composition in the drainage solution when the solution is reused in closed hydroponic cultivation of strawberries [126].

5.4 Utilization of renewable energy in agriculture

On the national level, the proportion of renewable energy is increasing due to the influence of RE100, etc. Farms are also installing small-scale solar power generation facilities to generate electricity and use it for environmental control. Since a lot of electric energy is needed for heating and cooling, the trend of increasing installation of small-scale renewable energy facilities is currently slowing down. This is because the perception has spread that it would be better not to use renewable energy at all, as replacing heating and cooling energy with renewable energy requires a lot of facility installation costs.

5.5 Introduction of medicinal crops

Hydroponics has been mainly used to improve the quality of vegetables such as lettuce, tomatoes, cucumbers, bell peppers, peppers, and melons. Recently, due to the diversity of consumers, the cultivation of medicinal plants with high economic value has also increased [127].

Cultivating medicinal crops with high demand under customized environmental conditions can achieve high yields and high quality, shorten the cultivation period, and enable sustainable cultivation [74].

Ice plant is known as a highly functional plant that is good for preventing adult diseases in Japan [128], and is also recognized as a high value-added crop in Korean plant factories. In Korea, plant factories are performing well in the niche market, not only carrot [13], leaf kale [19], *Peucedanum japonicum* [24], ragweed [76], ice plant [106], quinoa [129], but also Indian pennywort (*Centella asiatica*), which is used as a food and pharmaceutical material with rapidly increasing domestic demand [130, 131].

6. Conclusion

Consumers increasing interest in income, health, and the environment is already reflected in the market, and their preference for high-quality agricultural products is reflected.

Hydroponics facilitates year-round crop production by meeting consumers' demands for consistent and diverse agricultural products in a clean environment. It is necessary to promote to general consumers that hydroponics and plant factory cultivation are eco-friendly by minimizing irrigation and fertilizer amounts, preventing environmental pollution, and increasing space utilization.

In order to achieve eco-friendly cultivation, recycling technologies for media and nutrient solutions should be developed and expanded, and sales and profitability improvement through light efficiency and shortened growing periods are important for energy savings.

Agricultural products produced in plant factory systems should be price competitive with general agricultural products through the development of plant factory-specific species produced on-site, cultivation technologies, productivity improvements, and production cost reduction technologies.

The labor shortage will also act as a stimulating factor for the development of hydroponics, aquaponics, and plant factories, and automation element technologies for labor savings are expected to develop. Various control facilities combining hydroponics or plant factory technology with the Internet of Things (IoT), artificial intelligence (AI), and deep learning will be introduced.

We must continuously strive to develop quality and crisis response technologies for the stable production of high-quality agricultural products and consumers' interest in the environment.

Author details

Guangjae Lee[1]*, Jeongwook Heo[2] and Dongeok Kim[3]

1 LCM Science, Hwaseong-si, Republic of Korea

2 Rural Development Administration, Wanju-gun, Republic of Korea

3 Korea National College of Agriculture and Fisheries, Jeonju-si, Republic of Korea

*Address all correspondence to: ds3inj@naver.com

References

[1] Benke K, Tomkins B. Future food-production systems: Vertical farming and controlled-environment agriculture. Sustainability: Science, Practice and Policy. 2017;**13**:13-26

[2] Ting KC, Lin T, Davidson PC. Integrated urban controlled environment agriculture systems. In: LED Lighting for Urban Agriculture. Berlin, Germany: Springer; 2016. pp. 919-936

[3] Kaiser E, Correa Galvis V, Armbruster U. Efficient photosynthesis in dynamic light environments. A chloroplast's perspective. The Biochemical Journal. 2019;**476**:2725-2741

[4] Cerdán PD, Chory J. Regulation of flowering time by light quality. Nature. 2003;**423**:881-885

[5] Tripathi S, Hoang Q, Han Y, Kim J. Regulation of photomorphogenic development by plant phytochromes. International Journal of Molecular Sciences. 2019;**20**:6165

[6] Hangarter RP. Gravity, light and plant form. Plant, Cell & Environment. 1997;**20**:796-800. DOI: 10.1046/j.1365-3040.1997.d01-124.x

[7] Lee G, Yoon J, Lee M. Effects of irradiation quantities of mixed red and blue LEDs on the growth, total flavonoids and phenolic compound contents of leaf lettuce grown in plant factory. Acta Horticulturae. 2015;**1107**:245-250

[8] Cho JY, Son DM, Kim JM, Seo BS, Yang SY, Kim BW, et al. Effects of LEDs on the germination, growth and physiological activities of amaranth sprouts. Korean Journal of Horticultural Science and Technology. 2008;**26**:106-112

[9] Won J, Lee J, Cho B, Han C, Kang T. Growth characteristics, antixoident capacity and total phenolic content of lettuce according to generation position of air anion in a closed-type plant factory. Korean Journal of Horticultural Science and Technology. 2018;**36**:820-830. DOI: 10.12972/kjhst.20180080

[10] Cheigh H-S, Lee C-Y. Antioxidative charateristics of plant phenol compounds. Life Sciences. 1993;**3**:9-17

[11] Jin YX, Cho YS, Choi YM. Nutritional quality of *Peucedanum japonicum* Thunb. leaves in relation to ripening time, growing condition and blanching. Korean Journal of Food Preservation. 2014;**21**:784-789

[12] Lee GJ. Effects of red and blue light emitting diode (LED) irradiation on growth and yield in hydroponically grown leaf lettuce (*Lactuca sativa* L.). In: Annual Reports of Chungbuk Agricultural Research and Extension Service. Cheongju. Chungbuk. Vol. 2014. 2014. pp. 242-251

[13] Cho Y-Y, Cha M-K, Yang Gyu K, Kim HC, Bae JH. Effect of different culture nutrient solution EC on carrot top growth and nutritional contents in a closed–type plant factory system. Korean Journal of Horticultural Science and Technology. 2018;**36**(1):37-45. DOI: 10.12972/kjhst.20180005

[14] Pedras MSC, Yaya EE, Glawischnig E. The phytoalexins from cultivated and wild crucifers: Chemistry and biology. Natural Product Reports. 2011;**28**:1381-1405

[15] Schmidt S, Zietz M, Rohn S, Kroh LW, Krumbei A. Genotypic and climatic influences on the concentration

and composition of flavonoids in kale (*Brassica oleracea* var. sabellica). Food Chemistry. 2010;**119**:1293-1299

[16] Mithen R. Leaf glucosinolate profile and their relationship to pest and disease resistance in oilseed rape. Euphytica. 1992;**63**:71-83

[17] Kushad MM, Brown AF, Kurilich AC, Juvik JA, Klein BP, Wallig MA, et al. Variation of glucosinolates in vegetable crops of *Brassica oleracea*. Journal of Agricultural and Food Chemistry. 1999;**47**:1541-1548. DOI: 10.1021/jf980985s

[18] Lee MJ, Lim SY, Kim JK, Oh MM. Heat shock treatments induce the accumulation of phytochemicals in kale sprouts. Korean Journal of Horticultural Science and Technology. 2012;**30**:509-518

[19] Lee G, Heo J, Jung C, Kim H, Jo J, Lee J, et al. Effects of artificial light sources on growth and glucosinolate contents of hydroponically grown Kale in plant factory. Protected Horticulture and Plant Factory. 2016;**25**:77-82. DOI: 10.12791/KSBEC.2016.25.2.77

[20] Huseby S, Koprivova A, Lee BR, Saha S, Mithen R, Wold AB, et al. Diurnal and light regulation of sulphur assimilation and glucosinolate biosynthesis in Arabidopsis. Journal of Experimental Botany. 2013;**64**:1039-1048. DOI: 10.1093/jxb/ers378

[21] Steindal ALH, Molmann J, Bengtsson GB, Johansen TJ. Influence of daylength and temperature on the content of health-related compounds in Broccoli (*Brassica oleracea* L. var. italica). Agriculture and Food Chemistry. 2013;**61**:10779-10786. DOI: 10.1021/jf403466r

[22] Lefsrud MG, Kopsell DA, Sams CE. Irradiance from distinct wavelength light-emitting diodes affect secondary metabolites in kale. HortScience. 2008;**43**:2243-2244. DOI: 10.21273/HORTSCI.43.7.2243

[23] Choi J, Bok G, Lee H, Do H, Park J. Effect of red and blue LED ratio on growth and glucosinolate contents of watercress (*Nasturtium officinale*) in a plant factory. Korean Journal of Horticultural Science and Technology. 2020;**38**:474-486. DOI: 10.7235/HORT.2020004

[24] Lee G, Heo J, Jung C, Kim H, Yoon J, Kim D, et al. Effects of plant factory cultural systems on growth, vitamin c and amino acid contents, and yield in hydroponically grown *Peucedanum japonicum*. Protected Horticulture and Plant Factory. 2015;**24**:281-286. DOI: 10.12791/KSBEC.2015.24.4.281

[25] Samuoliene G, Sirtautas R, Brazaityte A, Viršilė A, Duchovskis P. Supplementary red LED lighting and the changes in phytochemical content of two baby leaf lettuce varieties during three seasons. Journal of Food, Agriculture and Environment. 2012;**10**:701-706

[26] Li H, Tang C, Xu Z, Liu X, Han X. Effects of different light sources on the growth of nonheading chinese cabbage (*Brassica campestris* L.). The Journal of Agricultural Science. 2012;**4**:262-273. DOI: 10.5539/jas.v4n4p262

[27] Kim Y, Lee J. Growth and contents of anthocyanins and ascorbic acid in lettuce as affected by supplemental UV-A LED irradiation with different light quality and photoperiod. Korean Journal of Horticultural Science and Technology. 2016;**34**:596-606. DOI: 10.12972/kjhst.20160061

[28] Lee MH. Changes in nutritive components by growth periods in spinach grown outdoors in autumn.

Journal of the East Asian Society of Dietary Life. 2009;**19**:169-179

[29] Brazaityte A, Viršilė A, Jankauskiene J, Sakalauskiene S, Samuoliene G, Sirtautas R, et al. Effect of supplemental UV-A irradiation in solid-state lighting on the growth and phytochemical content of microgreens. International Agrophysics. 2015;**29**:13-22. DOI: 10.1515/intag-2015-0004

[30] Kim YD, Choi SK. Studies on utilization of medicinal herbs as vegetable by hydroponics. 3. Component analysis of medicinal herbs as vegetable. Journal Oriental Botany Research. 1995;**8**:281-286

[31] Li Q, Kubota C. Effects of supplemental light quality on growth and phytochemicals of baby leaf lettuce. Environmental and Experimental Botany. 2009;**67**:59-64

[32] Ninu L, Ahmad M, Miarelli C, Cashmore AR, Giuliano G. Cryptochrome controls tomato development in response to blue light. The Plant Journal. 1999;**18**:551-556

[33] Giliberto L, Perrotta G, Pallara P, Weller JL, Fraser PD, Bramley PM, et al. Manipulation of the blue light photoreceptor cryptochrome in tomato affects vegetative development, flowering time, and fruit antioxidant content. Plant Physiology. 2005;**137**:199-208. DOI: 10.1104/pp.104.051987

[34] Reyes LF, Miller JC, Cisneros-Zevallos L. Environmental conditions influence the content and yield of anthocyanins and total phenolic in purple- and red-flesh potatoes during tuber development. American Journal of Potato Research. 2004;**81**:187-193. DOI: 10.1007/BF02871748

[35] Uleberg E, Rohloff J, Jaakola L, Trôst K, Junttila O, Haggman H. Effects of temperature and photoperiod on yield and chemical composition of northern and southern clones of bilberry (*Vaccinium myrtillus* L.). Journal of Agricultural and Food Chemistry. 2012;**60**:10406-10414. DOI: 10.1021/jf302924m

[36] Lee J, Kim Y. Growth and anthocyanins of lettuce grown under red or blue light-emitting diodes with distinct peak wavelength. Korean Journal of Horticultural Science and Technology. 2014;**32**(3):330-339. DOI: 10.7235/hort.2014.13152

[37] Hwang H, Jeong H, Jeong J, Jo D, Hwang S. Changes in growth and anthocyanin content of *Brassica juncea* L. affected by light intensity and photoperiod in plant factory with artificial lighting. Korean Journal of Horticultural Science and Technology. 2022;**40**(6):586-594. DOI: 10.7235/HORT.20220053

[38] Kaufmann C. Reducing tip burn in lettuce grown in an indoor vertical farm: Comparing the impact of vertically distributed airflow vs. horizontally distributed airflow in the growth of *Lactuca sativa* (order No. 30491903) [IOP publishing proQuest dissertations and theses global]. Available from: https://www.proquest.com/dissertations-theses/reducing-tipburn-lettuce-grown-indoor-vertical/docview/2825771585/se-2 [Accessed: September 25, 2024]

[39] Kubota C, Papio G, Ertle J. Technological overview of tipburn management for lettuce (*Lactuca sativa*) in vertical farming conditions. Acta Hortic. 1369. pp. 65-74. DOI: 10.17660/ActaHortic.2023.1369.8

[40] Bangerth F. A role for auxin and auxin transport inhibitors on the Ca content of artificially induced

parthenocarpic fruits. Physiologia Plantarum. 1976;**37**:191-194

[41] Chung HD. Leaf absorption and translocation of $^{45}CaCl_2$ in tomato (*Lycopersicon esculentum*) plants. Journal of the Korean Society for Horticultural Science. 2001;**42**:651-655

[42] Kreij CD, Van Goor VJ, Van Doesburg JD. The incidence of calcium oxalte crystals in fruit walls of tomato as affected by humidity, phosphate and calcium supply. Journal of Horticultural Sciences. 1992;**67**:45-50

[43] Lee G, Kim T. Effects of the use of additional calcium sources on growth and fruit characteristics in tomato hydroponics. Horticulture, Environment and Biotechnology. 2010;**51**:360-366

[44] Kang YK. The components and use methods for preventing physiological disorder of plants. Korean Intellectual Property Office. 1992. Konzession No. 10-0053370

[45] Hong JH, Lee SK. Effects of calcium treatment on tomato fruit ripening. Journal of the Korean Society for Horticultural Science. 1999;**40**:638-642

[46] Hong JH, Lee SK. Changes in cell wall materials of tomato fruits during ripening. Journal of the Korean Society for Horticultural Science. 1995;**36**:57-61

[47] Kumaratenna KPS, Cho Y. Detection of tipburn stress on lettuce grown in a plant factory using artificial intelligence (AI) models. Korean Journal of Horticultural Science and Technology. 2024;**42**(6):1-14. DOI: 10.7235/HORT.20240059

[48] Watkinson JH. Prevention of selenium deficiency in grazing animals by annual top dressing of pasture with sodium selenate. New Zealand Veterinary Journal. 1983;**31**:78-85

[49] Chen LC. Determination of selenium concentration of rice in China and effect of fertilization of selenite and selenate on selenium content of rice. Journal of Agricultural and Food Chemistry. 2002;**50**:5128-5130

[50] Hu QH. Effects of selenium on the yield and quality of green tea leaves harvested in early spring. Journal of Agricultural and Food Chemistry. 2003;**51**:3379-3381

[51] National Research Council (NRC). Recommended dietary allowances. In: 9th Food and Nutrition Board Committee on Dietary Allowances. Washington D.C.: National Academy; 1980

[52] Park KW, Lee JH, Geryer B. Effects of selenium concentrations in nutrient solution on the growth and contents of inorganic substances of Chinese leaf vegetables. Journal of the Korean Society for Horticultural Science. 1996;**37**:47-54

[53] Lee G, Kang B, Kim T, Kim T, Kim J. Effects of different selenium concentrations of the nutrient solution on the growth and quality of tomato fruit in hydroponics. Acta Horticulturae. 2007;**761**:443-448

[54] Benton J, Jr J. Hydroponics: History and use in plant nutrition studies. Journal of Plant Nutrition. 1982;**5**(8):1,003-1,030. DOI: 10.1080/01904168209363035

[55] Hydeoponic Market Out Look (2023 to 2024). 2024. Available from: https://www.futuremarketinsights.com/reports/hydroponics-market?utm_source=adwords&utm_medium=ppc&utm_source=adwords&utm_medium=ppc&id=5&gad_source=1&gclid=CjwKCAjw6c63BhAiEiwAF0EH

1IETzhutYDeUlAWAJzuUVRaaXRCV JyEFjfdOxuxzT9EtbKcV_1NFUBoC6Q wQAvD_BwE [Accessed: September 25, 2024]

[56] Kaiser C, Ernst M. Hydroponic Lettuce. University of Kentucky College of Agriculture, Food and Environment, Cooperative Extension Service. Center for Crop Diversification Crop Profile CCD-CP-63; 2012. Available from: https://publications.ca.uky.edu/sites/publications.ca.uky.edu/files/hydrolettuce.pdf

[57] Makhadmeh IM, Al-Tawaha A, Edaroyati P, Al-Karaki G, Al-Tawaha AR, Hassan SA. Effects of different growth media and planting densities on growth of lettuce grown in a closed soilless system. Research on Crops. 2017;**18**(2):294-298

[58] Al-Ajmi A, Al-Karaki G, Othman Y. Effect of different substrates on fruit yield and quality of cherry tomato grown in a closed soilless system. Acta Hortic. 2009;**807**:491-494. DOI: 10.17660/ActaHortic.2009.807.71

[59] Martinez HP. O uso do Cultivo Hidropônico de Plantas em Pesquisa. 2nd ed. Viçosa: UFV; 1999. p. 47. (Cadernos didáticos, 1)

[60] Garcia IP, Marques MC, Silva VTA. Olericultura, saladas menos convencionais. In: FNP Consultoria E COMÉRCIO. MAPA of Brazil. Agrianual-98. São Paulo, Brazil; 1997. pp. 31-36

[61] Altunlu H, Ayranci Y, Gercek S. Comparison of water pillow and drip irrigation systems for tomato (*Lycopersicon esculentum* Mill.) production under greenhouse conditions in the mediterranean region of Turkey. Bulgarian Journal of Agricultural Science. 2017;**23**:76-82

[62] Al-Tawaha AR, Al-Karaki G, Al-Tawaha AR, Sirajuddin SN, Makhadmeh I, Wahab PEM, et al. Effect of water flow rate on quantity and quality of lettuce (*Lactuca sativa* L.) in nutrient film technique (NFT) under hydroponics conditions. Bulgarian Journal of Agricultural Science. 2018;**24**:793-800

[63] Wheeler RM, Hinkle CR, Mackowiak CL, Sager JC, Knott WM. Potato growth and yield using nutrient film technique (NFT). American Potato Journal. 1990;**67**:177-187

[64] Ronan G, de Oliveira PSR, Resende FV. Long-life tomato cultivars growing under the hydroponic nutrient film technique. Scientia Agricola. 2002;**59**:803-806

[65] Baganz GF, Junge R, Portella MC, Goddek S, Keesman KJ, Baganz D, et al. The aquaponic principle—It is all about coupling. Reviews in Aquaculture. 2022;**14**:252-264. DOI: 10.1111/raq.12596

[66] Barbosa LG, Almeida GFD, Kublik N, Proctor A, Reichelm L, Weissinger E, et al. Comparison of land, water, and energy requirements of lettuce grown using hydroponic vs. conventional agricultural methods. International Journal of Environmental Research and Public Health. 2015;**12**:6879-6891. DOI: 10.3390/ijerph120606879

[67] Chen JM, Ma JF. Abiotic and biological mechanisms of nitric oxide removal from waste air in biotrickling filters. Journal of the Air & Waste Management Association (1995). 2006;**56**:32-36. DOI: 10.1080/10473289.2006.10464433

[68] Schreier HJ, Mirzoyan N, Saito K. Microbial diversity of biological filters in recirculating aquaculture systems. Current Opinion in Biotechnology.

2010;**21**:318-325. DOI: 10.1016/j.copbio.2010.03.011

[69] Choi K, Kim K, Shawon MRA, Lee H, Yoon Y. Effect of nutrition supplement on water quality and growth of geranium in aquaponic system. Korean Journal of Horticultural Science and Technology. 2024;**43**(1):1-15. DOI: 10.7235/HORT.20250012

[70] Heo JW, Lee YB, Kim DE, Chang YS, Chun CH. Effects of supplementary LED lighting on growth and biochemical parameters in Dieffenbachia amoena 'Camella' and Ficus elastica 'Melany'. Korean Journal of Horticultural Science and Technology. 2010;**28**:51-58

[71] Kozai T. Smart Plant Factory: The Next Generation Indoor Vertical Farms. Singapore: Springer; 2018. p. 238. DOI: 10.1007/978-981-13-1065-2

[72] Mills E. The carbon footprint of indoor cannabis production. Energy Policy. 2012;**46**:58-67

[73] Ikeda A, Tanimura Y, Ezaki K, Kawai Y, Nakayama S, Iwao K, et al. Environmental control and operation monitoring in a plant factory using artificial light. Acta Horticulturae. 1992;**304**:151-158. DOI: 10.17660/ActaHortic.1992.304.16

[74] Park JE, Kim H, Kim J, Choi SJ, Ham J, Nho CW, et al. A comparative study of ginseng berry production in a vertical farm and an open field. Industrial Crops and Products. 2019;**140**:111612. DOI: 10.1016/j.indcrop.2019.111612

[75] Kang MH, Han YM, Lee YJ, Ji Y, Seong HA, Kim HD. Seasonal variations in bioactive compounds and antioxidant activity in the leaves of *Sasa quelpaertensis*. Horticultural Science. 2023;**41**:403-413. DOI: 10.7235/HORT.20230037

[76] Qin L, Guo S, Ai W, Tang Y. Selection of candidate salad vegetables for controlled ecological life support system. Advances in Space Research. 2008;**41**:768-772

[77] Wicharuck S, Khongdee N, Pripanakul T, Whangchai N, Pimpimol T, Chaichana C. Vertical farming: A potential farming practice for lettuce production. Chilean Journal of Agricultural Research. 2023;**83**:248-259. DOI: 10.4067/S0718-58392023000300248

[78] Gaudreau L, Charbonneau J, Vézina LP, Gosselin A. Photoperiod and photosynthetic photon flux influence growth and quality of greenhouse-grown lettuce. HortScience. 1994;**29**:1285-1289. DOI: 10.21273/HORTSCI.29.11.1285

[79] Heo J, Kim H, Lee K, Yoon J, Lee J, Huh Y, et al. Effect of supplementary radiation on growth of greenhouse-grown kales. Korean Journal of Environmental Agriculture. 2015;**34**:38-45. DOI: 10.5338/KJEA.2015.34.1.02

[80] Hogewoning SW, Douwstra P, Trouwborst G, van Ieperen W, Harbinson J. An artificial solar spectrum substantially alters plant development compared with usual climate room irradiance spectra. Journal of Experimental Botany. 2010;**61**:1267-1276

[81] Kim HH, Goins GD, Wheeler RM, Sager JC. Green-light supplementation for enhanced lettuce growth under red- and blue-light-emitting diodes. HortScience. 2004;**39**:1617-1622

[82] Velez-Ramirez A, Heuvelink E, van Ieperen W, Vreugdenhil D, Millenaar F. Continuous light as a way to increase greenhouse tomato production: Expected

challenges. Acta Hortic. 2015;**956**:51-57. DOI: 10.17660/ActaHor-tic.2012.956.3

[83] Schuerger AC, Brown CS, Stryjewski EC. Anatomical features of pepper plants (*Capsicum annuum* L.) grown under red light-emitting diodes supplemented with blue or far-red light. Annals of Botany. 1997;**79**:273-282. DOI: 10.1006/anbo.1996.0341

[84] Carvalho RF, Takaki M, Azevedo RA. Plant pigments: The many face of light perception. Acta Physiologiae Plantarum. 2011;**33**:241-248. DOI: 10.1007/s11738-010-0533-7

[85] Kozai T, Fujiwara K, Runkle ES. LED Lighting for Urban Agriculture. Singapore: Springer Science and Business Media; 2016. DOI: 10.1007/978-981-10-1848-0

[86] Lin KH, Huang MY, Huang WD, Hsu MH, Yang ZW, Yang CM. The effects of red, blue, and white light-emitting diodes on the growth, development, and edible quality of hydroponically grown lettuce (*Lactuca sativa* L. var. capitata). Scientia Horticulturae. 2013;**150**:86-91

[87] Lee HI, Kim YH. Utilization efficiencies of electric energy and photosynthetically active radiation of lettuce grown under red LED, blue LED and fluorescent lamps with different photoperiods. Journal of Biosystems Engineering. 2013;**8**:279-286. DOI: 10.5307/JBE.2013.38.4.279

[88] Dorais M, Gosselin A, Trudel MJ. Annual greenhouse tomato production under a sequential intercropping system using supplemental light. Scientia Horticulturae. 1991;**45**:225-234. DOI: 10.1016/0304-4238(91)90067-9

[89] An CG, Hwang YH, An JU, Yoon HS, Chang YH, Shon GM, et al. Effect of LEDs (light emitting diodes) irradiation on growth of paprika (*Capsicum annuum* 'Cupra'). Journal of Bio-Environment Control. 2011;**20**:253-257

[90] Johkan M, Shoji K, Goto F, Hashida S, Yoshihara T. Blue light-emitting diode light irradiation of seedlings improves seedling quality and growth after transplanting in red leaf lettuce. HortScience. 2010;**45**:1809-1814

[91] Xu H, Xu Q, Li F, Feng Y, Qin F, Fang W. Applications of xerophytophysiology in plant production-LED blue light as a stimulus improved the tomato crop. Scientia Horticulturae. 2012;**148**:190-196

[92] Choi MK, Baek GY, Kwon SJ, Yoon YC, Kim HT. Effect of LED light wavelength on lettuce growth, vitamin C and anthocyanin contents. Protected Horticulture and Plant Factory. 2014;**23**:19-25. DOI: 10.12791/KSBEC.2014.23.1.019

[93] Moan J. Visible light and UV radiation. In: Brune D, Helborg R, Persson BRR, Paakkonen R, editors. Radiation at Home, Outdoors and in the Workplace. Oslo, Norway: Scandinavian Science Publishers; 2001

[94] Cho KJ, Cho JY, Park IS, Oh W. Effects of duty ratio of pulsed LED light on growth and photosynthetic rate of lettuce grown in a plant factory system. Journal of People Plants and Environment. 2013;**16**:472-434. DOI: 10.11628/ksppe.2013.16.6.427

[95] Dong C, Shao L, Liu G, Wang M, Liu H, Xie B, et al. Photosynthetic characteristics, antioxidant capacity and biomass yield of wheat exposed to intermittent light irradiation with millisecond-scale periods. Journal of Plant Physiology. 2015;**184**:28-36. DOI: 10.1016/j.jplph.2015.06.012

[96] Mori Y, Takatsuji M, Yasuoka T. Effects of pulsed white LED light on the growth of lettuce. Journal of Society of High Technology in Agriculture. 2002;**14**:136-140

[97] Son K, Lee S, Myungmin O. Comparison of lettuce growth under continuous and pulsed irradiation using light-emitting diodes. Korean Journal of Horticultural Science and Technology. 2018;**36**:542-551. DOI: 10.12972/kjhst.20180054

[98] Pardo GP, Velázquez ST, Orea AC, Aguilar CH, Pacheco FAD, Martínez FR. Pulsed LED light in germination and growth of lettuce seeds. Bothalia Journal. 2016;**46**:13-26

[99] Salisbury FB, Ross CW. Plant Physiology. 4th ed. Belmont, CA, USA: Wadsworth; 1992. pp. 27-65

[100] Hunter DC, Burritt DJ. Light quality influences adventitious shoot production from cotyledon explants of lettuce (*Lactuca sativa* L.). In Vitro Cellular & Developmental Biology - Plant. 2004;**40**:215-220

[101] Graves CJ. The nutrient film technique. Horticultural Reviews. 1983;**5**:1-44. DOI: 10.1002/9781118060728.ch1

[102] Nemali K. Details of Electrical Conductivity Measurements in Greenhouse Production. Purdue: Purdue University; 2018

[103] Vande Vooren J, Welles G, Hayman G. Glasshouse crop production. In: The Tomato Crop. Vol. 5. Dordrecht, The Netherlands: Springer; 1986. pp. 581-623

[104] Samarakoon UC, Weerasinghe PA, Weerakkody WAP. Effect of electrical conductivity (EC) of the nutrient solution on nutrient uptake. Growth, and yield of leaf lettuce (*Lactuca sativa* L.) in stationary culture. Tropical Agricultural Research. 2006;**18**:13-21

[105] Tao L, Hongjun Y, Wang T, Zhang T, Shi C, Jiang W. Influence of the electrical conductivity of the nutrient solution in different phenological stages on the growth and yield of cherry tomato. Horticulturae. 2022;**8**(5):378.(1-14). DOI: 10.3390/horticulturae8050378

[106] Cha M, Kim J, Cho Y. Growth model of common ice plant (*Mesembryanthemum crystallinum* L.) using expolinear functions in a closed-type plant production system. Korean Journal of Horticultural Science and Technology. 2014;**32**:493-498. DOI: 10.7235/hort.2014.14013

[107] Jensen MH. Hydroponics. HortScience. 1997;**32**:1018-1021

[108] Nederhoff E, Stanghellin C. Water use efficiency of tomatoes. Practical Hydroponics and Greenhouses. 2010;**115**:52

[109] Savvas D, Sigrimis N, Chatzieustratiou E, Paschalidis C. Impact of a progressive Na and Cl accumulation in the root zone on pepper grown in a closed-cycle hydroponic system. Acta Horticulturae. 2009;**807**:451-456

[110] Allende A, Monaghan L. Irrigation water quality for leafy crops: A perspective of risks and potential solutions. International Journal of Environmental Research and Public Health. 2015;**12**:7457-7477. DOI: 10.3390/ijerph120707457

[111] Zekki H, Gauthier L, Gosselin A. Growth, productivity, and mineral composition of hydroponically cultivated greenhouse tomatoes, with or without nutrient solution recycling. Journal of

the American Society for Horticultural Science. 1996;**121**:1082-1088. DOI: 10.21273/jashs.121.6.1082

[112] Miller A, Adhikari R, Nemali K. Recycling nutrient solution can reduce growth due to nutrient deficiencies in hydroponic production. Frontiers in Plant Science. 2020;**11**:607643. DOI: 10.3389/fpls.2020.607643

[113] Ko MT, Ahn T, Cho Y, Son J. Uptake of nutrients and water by paprika (*Capsicum annuum* L.) as affected by renewal period of recycled nutrient solution in closed soilless culture. Protected Horticulture. 2013;**54**:412-421

[114] Brun R, Settembrino A, Couve C. Recyling of nutrient solutions for rose (*Rosa hybrida*) in soiless culture. Acta Horticulturae. 2001;**554**:183-192. DOI: 10.17660/ActaHortic.2001.554.19

[115] Lykas C, Katsoulas N, Giaglaras P, Kittas C. Electrical conductivity and pH prediction in a recirculated nutrient solution of a greenhouse soilless rose crop. Journal of Plant Nutrition. 2006;**29**:1585-1599. DOI: 10.1080/01904160600848904

[116] Sonneveld C, Voogt W. Plant Nutrition of Greenhouse Crops. New York, NY, USA: Springer; 2009. DOI: 10.1007/978-90-481-2532-6

[117] Lee J, Shin J. Comparison of basal physicochemical changes of reused rockwool substrate in hydroponic tomato cultivation. Journal of Bio-Environment Control. 2024;**33**:114-119. DOI: 10.12791/KSBEC.2024.33.2.114

[118] Agius C, Von Tucher S, Rozhon W. The effect of salinity on fruit quality and yield of cherry tomatoes. Horticulturae. 2022;**8**:59

[119] Kuncoro CBD, Asyikin MBZ, Amaris A. Development of an Automation System for Nutrient Film Technique Hydroponic Environment. Vol. 207. Advances in Engineering Research; 2021. pp. 437-443. DOI: 10.2991/aer.k.211106.069

[120] Putra PA, Yuliando H. Soilless culture system to support water use efficiency and product quality: A review. Agriculture and Agricultural Science Procedia. 2015;**3**:283-288

[121] Sun J, Gao HB, Tian J, Wang JW, Du CX, Guo SR. Development status and trends of protected horticulture in China. Journal of Nanjing Agricultural University. 2019;**42**:594-604

[122] Bohme M. Effect of closed systems in substrate culture for vegetable production in greenhouses. Acta Horticulturae. 1995;**396**:45-54. DOI: 10.17660/ActaHortic.1995.396.4

[123] Raviv M, Lieth JM. Soilless Culture: Theory and Practice. New York, USA: Elsevier; 2008

[124] Sakamoto M, Suzuki T. Effect of nutrient solution concentration on the growth of hydroponic sweet potato. Agronomy. 2020;**10**:1708

[125] Jones JB. Hydroponics: Its history and use in plant nutrition studies. Journal of Plant Nutrition. 1982;**5**:1003-1030

[126] Lee H, Kim H, Jang W, Nam M, Lee I, Lee M, et al. Changes in chemical characteristics of drainage solution as influenced by years of root media use in successive hydroponic cultivations of 'Seolhyang' strawberries. Korean Journal of Horticultural Science and Technology. 2019;**37**:375-383. DOI: 10.7235/HORT.20190038

[127] Al-Tawaha A, Al-Karaki G, Massadeh A. Effects of planting density and cutting height on herbage and water use efficiency of thyme (*Origanum syriacum* L.) grown under protected soilless and open fi eld conditions. Research on. Crops. 2016;**17**:118-128

[128] Agarie S, Kawaguchi A, Kodera A, Sunagawa H, Kojima H, Nose A, et al. Potential of the common ice plant, *Mesembryanthemum crystallinum* as a new highfunctional food as evaluated by polyol accumulation. Plant Production Science. 2009;**12**:37-46

[129] Austin J, Jeon YA, Cha M-K, Park S, Cho Y-Y. Effects of photoperiod, light intensity and electrical conductivity on the growth and yield of quinoa (*Chenopodium quinoa* Willd.) in a closedtype plant factory system. Korean Journal of Horticultural Science and Technology. 2016;**34**(3):405-413

[130] Bylka W, Znajdek-Awiżeń P, Studzińska-Sroka E, Brzezińska M. *Centella asiatica* in cosmetology. Postepy Dermatologii I Alergologii. 2013;**30**:46-49

[131] Ha JH, Kwon MC, Kim SS, Jeong MH, Hwang B, Lee HY. Enhancement of skin-whitening and UV-protective effects of *Centella asiatica* L. urban by ultrasonification process. The Korean Journal of Medicinal Crop Science. 2010;**18**:79-85

Chapter 4

Perspective Chapter: An Overview of Hydroponic Cultivation for Sustainable Food Production

Bonga Lewis Ngcobo, Nokuzola Phungula, Philile Ngcobo and Zandile Maninjwa

Abstract

Global food security is increasingly challenged by unpredictable climatic conditions and population growth. Currently, most farmers rely on soil-based cultivation methods for food production. The limitations of this approach mainly include high dependence on the seasonal changes and chemical additives. These limitations suggest that traditional cultivation methods may not be sufficient to supply the world's food needs in the future. As a result, alternative, sustainable food production methods are needed. Hydroponic technology has emerged as a promising alternative, allowing for improved food production at both local and commercial scales. This review article, therefore, explores the potential of hydroponic systems to support plant growth and further looks at the performance of various crops in hydroponic systems. The key findings from the literature point out that while lettuce is still a common food crop produced hydroponically, herbs, certain fruits and medicinal plants are also gaining popularity. The review also exposed a gap in the research regarding the impact of hydroponic systems on health-promoting compounds and secondary metabolites on plant species. In addition, the review provides evidence that hydroponic cultivation accelerates plant growth as compared to soil-based cultivation methods. Finally, the review highlights the role of technology in optimizing hydroponic practices.

Keywords: sustainability, climate change, food security, economy, controlled environment agriculture

1. Introduction

Food security is a major challenge worldwide, and this challenge is attributed to climate change and rapid population growth. The latter influences food demand significantly, as the population continues to grow the food demand also grows [1]. A projection made by United Nations indicates that by 2037, there will be 9 billion people on earth and 10 billion around 2050, to match this increase, a rise in global food production is required [2]. Currently, food production is dependent on soil-based cultivation method, and this method has several limitations, such as climate change, urbanization, pollution and seasonal variations. In addition, farmers rely mainly on chemical inputs to enhance food production. This approach affects soil health and is unfriendly to the environment [3]. Based on this, traditional cultivation method is not guaranteed to produce enough food in the future [1]. There is therefore a need to search for alternative methods to help enhance or improve food production sustainably. Recently, hydroponic technology has emerged as a promising and innovative method to improve food production at both local and commercial scales [4–6]. The nutrient solution in hydroponic systems is independent of substrates, such as perlite, sand, peat moss, etc. [6]. As compared to traditional cultivation methods, hydroponics is increasingly gaining popularity in agricultural sector due to several advantages it offers. To mention a few, it uses water efficiently, has improved control over nutrient levels, is able to produce plants in areas with poor soil quality and enables year-round cultivation, especially in space-constrained regions [6–8]. In addition, beyond cultivating food crops, specialty crops like herbs and ornamentals can be cultivated in hydroponics [9]. According to Gashgari, et al. [1], hydroponic cultivation has been proven to enhance plant growth as compared to the traditional cultivation system. Similarly, Mariyappillai et al. [10] revealed that hydroponically grown plants, such as lettuce (*Lactuca sativa*), grow quicker and result in more yield as compared to soil-dependent/grown plants. Furthermore, research by Groenveld et al. [11] and Goddek et al. [12] demonstrates that hydroponic cultivation promotes faster growth and results in an improved yield.

Overall, hydroponics offers a sustainable solution to meet the global food demand using fewer resources [13, 14]. Hydroponics also reduces land and water requirements by 75 and 90%, respectively, presenting a potential solution to global food scarcity [15, 16]. Based on this, some farmers are transitioning from traditional cultivation methods to hydroponic cultivation, intending to address global food security, which is threatened by climate change [17]. This literature review, therefore, provides a comprehensive discussion of hydroponic cultivation systems and their potential to enhance food production. The review further compares hydroponic cultivation systems to conventional soil-based methods. It also highlights the economic implications of hydroponic systems and lastly, the use of technology, such as light-emitting diodes (LEDs), in hydroponic system is highlighted.

2. Overview of the commonly used hydroponic systems in agriculture

There are various types of hydroponic systems that the growers can choose from, and each system is accompanied by its own advantages and limitations. These hydroponic systems are highlighted in **Table 1.**

2.1 The nutrient film technique (NFT)

The NFT is a closed hydroponic system with a simple design, and it is commonly used by both small-scale and commercial growers who specialize in the production of fast-growing crops, such as lettuce, baby greens and herbs [24]. In a study conducted on five lettuce varieties, NFT and EBB and flow systems were compared, the number of days to maturity was reduced by the NFT compared to the EBB and flow [25].

Hydroponic system	Benefits	Limitations
Nutrient film technique (NFT)	The recycling of water in the NFT allows the user to save water [18]	The continuous irrigation is energy-intensive [19]
Deep water culture	Uses a minimum initial investment [20]	There is little room for automation, and this system requires constant supervision [20]
Aeroponics	The use of aeroponics can save the grower about 98% of water, 60% of fertilizer and 100% on the cost of pesticides [21]	The possibility of blocked nozzles can have a negative impact on cultivated plants [20]
Wick system	The closed nature of wick system limits run-off and promotes proper plant nutrition, and this promotes the uniform production [22]	The wick system is suitable for small-scale farming and only accommodates plants with short life cycle [23]

Table 1.
The benefits and limitations associated with the use of various hydroponic cultivation methods.

Another recent study was conducted on spinach and lettuce grown in the NFT, which both exhibited high yield and high content of mineral nutrients [26]. NFT consists of grow channels powered by mechanical pumps to circulate nutrient solution making them available to plants [20]. In this system, plant roots are suspended in the gullies with a constant flow of nutrient solution necessary for growth of plants [27].

2.2 Deep water culture (DWC)

Deep water culture, also referred to as direct hydroponic culture, is a stationary hydroponic system consisting of floating rafts placed over containers filled with nutrient-rich solution [20]. In this growing system, plant roots are continuously immersed in nutrient-rich water solution [15]. This system is highly efficient, cost-effective and is admired for the cultivation of large quantities of lettuce varieties and other leafy green vegetables [28]. A study by Majid, et al. [29] recommends deep water culture for the cultivation of lettuce and other leafy vegetables. It is important to note that this technique can be used by both well-established and small-scale farmers, and requires constant aeration to increase the amount of dissolved oxygen, which is usually achieved with the aid of air bubblers [20]. The lack of oxygen can hinder the roots from functioning properly [30]. Continuous monitoring of not only the nutrient concentration and oxygen, but also the buildup of mold, fungi, water pH and electrical conductivity (EC), is important for proper growth and development of the plants [18].

2.3 Aeroponics

Another common hydroponic system is aeroponics, and this is a system that does not use any growing medium [21]. In this system, the plants are supported by either vertical or horizontal panels and their roots are misted by nutrient-rich solution [20]. Plants grow well in aeroponics because of the continuous exposure of the roots to the air, which promotes root aeration and further ensures that plants have full access to oxygen [31]. This system is mostly intended for small horticultural commodities, it necessitates a high capital investment and operational costs [20] and is ideal for growing leafy vegetables, such as lettuce and spinach (*Spinacia oleracea)* [18]. A study by Chandra et al. [32] demonstrates an increase in yield and total phenolics on leafy vegetables, herbs (basil, kale and parsley) and fruit crops (bell pepper, cherry tomato and squash) grown in the aeroponic system. This study compared the aeroponic system to the soil-based growing method. Another study revealed that aeroponics are also popular for potato tuber cultivation, as they can result in an improved yield per unit area compared to conventional tuber production method [33].

2.4 The wick system

The wick hydroponic system, also known as the passive or self-feeding system, is a no pump [34], aerator or electricity system [18]. It is best suitable for growing herbs where the plants are grown in a pot containing an inert medium and a wick (cloth), which sucks up the nutrient solution from its tank onto the plant [35]. The wick system is not compatible for plants with high water requirements [23]. There is limited or no research on this type of hydroponic system. Further studies need to focus on evaluating the response of plant species to this type of hydroponic system.

2.5 Vertical farming in the controlled environment

Vertical farming is a farming practice that takes place in a controlled environment where plants are stacked on top of each other in vertical racks [36]. This is a highly efficient system, and it produces a high yield because plants are stacked upwards (mostly in shelves against the wall) in layers maximizing the production area [37, 38]. The loss of agricultural land to urbanization and the increase of food demand have popularized vertical farming [36] compared to the traditional farming method. That is due to vertical farming only requiring 5% of water for the same output of produce grown in the open field [38]. A lot of horticultural crops can be grown in a vertical farm, including crops, such as microgreens, salad leaves, strawberries (*Fragaria species*), spinach (*Spinacia olaracea*) and culinary herbs [39]. Considering the land constraints especially in the urban areas, vertical farming is the efficient alternative way to grow crops in the hydroponic settings; this system, however, requires continuous controlling and monitoring of the temperature, light, relative humidity and nutrient levels [40].

3. Performance of agricultural commodities in the hydroponic systems

As explained above, hydroponics can be used to grow a variety of plants, including fruits, vegetables, flowers and medicinal plants [41]. This growing method allows farmers or producers to produce crops all year round using less area and achieve higher crop yields [9]. Several studies demonstrated that hydroponics minimizes environmental degradation, enhances food production and most importantly improves food sustainability [9, 42]. Hydroponically grown crops have a better nutritional content than soil-grown plants, leading to improved human health [43]. On the other hand, hydroponic systems have been known for growing fruit and vegetable crops such as lettuce. It has, however, been discovered that the number of agricultural commodities that are successfully grown in such systems is increasing, and this is due to the reliability and cost-effectiveness of this growing method. Recently, hydroponic systems, such as deep water culture and nutrient film technique, have been able to successfully grow food-based commodities, including cereals, fruit and vegetables as well as herbs (**Table 2**) [25, 29, 45, 47, 55]. It is unusual and surprising that medicinal plants and ornamental crops are also grown in hydroponic systems. **Table 2** shows the list of crops that can be grown in the hydroponic systems. Furthermore, various studies have compared the performance of crops grown conventionally and in hydroponics and discovered that growing crops in the latter systems have a greater potential to address global food security because they enhance yield of various crops, such as tomato [56], strawberry ('Ozark Beauty,' Fragaria x ananassa) [57] and lettuce [29], and also shorten time to maturity [29, 56]. A detailed comparison of crops grown conventionally and in the hydroponic system is presented in **Table 3**.

Name of crop	Type of hydroponic system	Reference
Cereals		
Maize *(Zea mays* cv. Kaveri-244+, cv. BHARAT Hybrid Sultan 702, cv. Getco seeds GP-901, cv. Essence-Platinum)	Deep water culture	[25]
Durum wheat (*Triticum turgidum* L. cv. Claudio W.)	NS	[44]
Maize *(Zea mays* L. var. Jubilee)	Deep water culture	[25]
Fruit and vegetable crops		
Tomato *(Solanum lycopersicum* L. cv. Bush Goliath)	NS	[45]
Tomato *(Solanum lycopersicum* L. cv. Pannovy*)*	Nutrient film technique	[45]
Pepper *(Capsicum annuum* L. cv. Friariello)	Nutrient film technique	[46]
Lettuce *(Lactuca sativa)*	Nutrient film technique	[47]
Lettuce *(Lactusa sativa* L., var. Longifolia)	Deep water culture and nutrient film technique	[29]
Chili (Capsicum annuum)	Deep water culture	[48]
Herbs		
Basil (Basil ocimum sp.)	Deep flow culture/deep water culture	[49]
Dill (Anethum graveolens L.*), Rocket (Eruca sativa), Coriander (Coriandrum sativum* L.*)* and *Parsley (Petroselinum crispum*)	Nutrient film technique	[47]
Ornamental crops		
Chrysanthemum (*Chrysanthemum × morifolium* Ramat.)	Deep water culture	[50]
French marigold (*Tagetes patula* L.) Marigold *(Tagetes patula)*	Deep flow technique	[51]
European black nightshade *(Solanum nigrum)* Marvel of Peru *(Mirabilis jalapa)*	NS	[52]
Chrysanthemums *(Chrysanthemum indicum)*	NS	[50]
Chrysanthemums *Chrysanthemum* cv. "Rajkumari"	NS	[53]
Medicinal plants		
Milkweed *(Euphorbia peplus)*	Deep water culture	[54]

Table 2.
Growth of various plant species in various hydroponic systems.

Crop	Yield /plant		Root length		Plant height (mm)		Leaf length (mm)		Number of fruits		Fruit weight (g)		Days for maturity		Source
	H	C	H	C	H	C	H	C	H	C	H	C	H	C	
Tomato	820	480							64	38			58	62	[56]
Giant lettuce	455.9	53 ± 5.8	33.1 ± 3.1	23 ± 2.3	—	—	—	—	—	—	—	—	—	—	[58]
Lettuce	1.64	1.16	—	—	—	—	—	—	—	—	—	—	40.5	48	[29]
Tomato	401.3	379.4	—	—	—	—	—	—	4	4	—	—	—	—	[59]
Strawberry	85	70	—	—	—	—	—	—	—	—	—	—	—	—	[57]
Cucumber	—	—	—	—	190	94	146		—	—	—	—	—	—	[1]
Squash	1249.9	836.17	—	—	—		—		6	5	208.3	167.23	—	—	[60]
Lettuce	1.64	1.160	—	—	—		—	105	—	—	—	—	40	48	[29]

Table 3.
Comparison of growth attributes and yield between hydroponically (H) and conventionally (C) grown agricultural commodities.

4. Economic viability of the hydroponics

From the economic point of view, hydroponic cultivation is cheaper than the traditional cultivation methods. Hydroponic systems are technically efficient when it comes to saving water and fertilizers [6]. There are, however, a few limitations with the hydroponic systems such as high startup costs [61]. Research studies have shown that the costs of producing hydroponic vegetables inside greenhouses are higher compared to traditional farming systems. The hydroponic systems require electricity as a source of energy, and costly equipment, such as air and water pumps [62, 63]. During the installation of the climate-controlled system, several aspects should be considered, including system availability, system efficiency, transportation cost and the initial cost of the system [64]. To overcome these challenges, the Kratky hydroponic system should be adopted. This system offers an opportunity to grow crops sustainably and it reduces the costs that are associated with advanced hydroponic systems. In addition, this method does not necessitate the use of timers, air pumps, climate monitoring systems or additional labour [65]. It, however, encourages the rapid growth of minor crops, such as lettuce and herbs, because they can be cultivated with just one initial application of the nutrient solution throughout the full cropping season [66]. In addition, it can boost food security while offering urban farmers and residents a profitable and sustainable business opportunity [63]. On the other side, Folorunso, et al. [67] investigated the profitability of hydroponics under two production scales, namely small- and medium scales, and this author discovered that at both production scales, hydroponic systems were found to be largely profitable. According to this research, small-scale farming produced a net return of R115.66/m^2 annually, while medium-scale farming produced a net return of R540.05/m^2 annually [67].

5. Environmental benefits of hydroponic systems

5.1 Climate change and water use efficiency

Climate change has become increasingly recognized as one of the greatest challenges to humankind and all other life on earth [68]. The agricultural industry is affected by extreme weather occurrences, such as heatwaves, droughts, floods and irregular precipitation [69, 70]. These climatic shifts have a significant impact on crop yield directly endangering the availability of food and nutrition [71]. Climate change has significant socioeconomic impacts in addition to its biological impacts. The aforementioned problems associated with climate change lead to reduced agricultural production, which in turn influences higher food costs [72]. To address these challenges, extensive adaptation and mitigation methods are required. Recently, hydroponics emerged as an effective and efficient method of food production that can reduce the pressure faced by farmers. Crops grown in hydroponics are not dependent on external environmental conditions. A controlled environment protects crops from external factors and ensures continuous production while minimizing the risks of crop loss. This therefore ensures that there is a continuous food supply for everyone [73].

In addition, climate change affects the water resources needed for agricultural productivity, as higher temperatures, increased evapotranspiration and decreased precipitation lead to a greater requirement for irrigation [74]. Water security is another common challenge worldwide, and limited access to clean water is getting worse day by day [75]. Seventy per cent of the water extracted from water bodies, globally, is used for agricultural production [76]. Soil farming consumes substantially more water than hydroponics, as much of the water is lost to leaching and not absorbed by the plant's roots [77]. The hydroponic cultivation alone saves about 95% of irrigation water [78]. In the hydroponic systems, water is not wasted because the plant is grown directly in the nutrient solution and water flows through the pipes where it can be recycled and reused [4]. To ensure that there is enough food supply, it is vital to adopt various innovative and climate smart agricultural practices such as hydroponics [79].

5.2 Carbon footprint reduction

Food production and consumption have a substantial environmental impact, and the current view is that we need to switch to more sustainable growing techniques [80, 81]. Conventional agriculture requires significant inputs of pesticides, herbicides, fertilizers and chemicals-based substances, which harm the soil and pose serious threats to human health and the environment [82]. Hydroponics is one of the sustainable growing techniques that is proven to reduce the risks associated with extreme weather and pests. It decreases reliance on toxic chemicals like insecticides and pesticides [83]. Hydroponic farming is a useful environmentally friendly technology that results in minimal negative effects on the environment [84] while enhancing growth and yield [85]. Hydroponics is an alternative to conventional agriculture and can boost output without producing a large amount of waste substances into the environment [82, 86]. Conventional agriculture depends primarily on soil and the soil becomes less fertile because of continuous, repeated cultivation, which also results in decreased productivity and quality attributes [87]. In contrast, hydroponics is a soilless method that ensures efficient land use [10]. Hydroponics allows more precise control over crop nutrition, resulting in more effective nutrient regulation and water management [88]. In hydroponic system, untreated wastewater can be used to irrigate crops, and this reduces the demand on natural water sources [89, 90]. Hydroponic systems can be a solution to agricultural food production challenges since they have a lower environmental impact, use less energy, water and fertilizer, and require less acreage while improving agricultural yield [91].

6. The use of technology to enhance production in hydroponics

Since the climate change and a rapid increase in the human population threaten the agricultural food systems, some farmers have slowly shifted to controlled environment agriculture (CEA) where limited space is used to sustainably grow high-quality crops. Hydroponic cultivation is one of the innovative and promising methods to grow crops, providing numerous advantages over conventional farming methods. Various crops have recently been produced successfully in hydroponic systems, as outlined in **Table 1**, and high nutritional quality and yield have been achieved, thus promoting sustainable development goals (SDGs) (**Figure 1**).

Most importantly, in CEA the use of solid-state light sources is the key technology to enhance photosynthetic activities [92]. Over the past years, light-emitting diodes (LEDs) have been preferred because of their unique properties and added advantages over other light sources, such as inflorescence and incandescence [93, 94]. Specifically, LEDs are cost effective, use less energy and have a long lifespan. In a recent review compiled by Ngcobo and Bertling [94], the application of LED

Figure 1.
Sustainable development goals that are directly affected by hydroponic cultivation (SDG 1- Zero hunger; SDG 2- No poverty; SDG 3- Good health and well-being; and SDG 13- Climate action).

technology in hydroponic systems is discussed briefly and it is highlighted that it plays a pivotal role in enhancing photosynthetic activities, thereby enhancing growth, yield and quality. Other authors have also proven that technology enhances yield and quality of various horticultural crops grown hydroponically [52, 95, 96]. LEDs affect plant morphological and physiological responses by manipulating photosynthesis as they allow wavelengths to be matched to those of plant photoreceptors, such as phytochrome red and far red. Additionally, different wavelengths of light affect photosynthetic pigments in plants, such as chlorophylls and carotenoids [97]. Without supplemental lights in hydroponics, photosynthesis cannot be optimized, as such the yield and quality can be compromised.

Through the Fourth Industrial Revolution, traditional farming has been transformed into a more efficient growing system through advancements in agricultural technology. As a result, smart hydroponic systems have been introduced. This growing technique involves controlling the operations systematically to enhance agricultural production [98]. This includes, amongst others, operations like automated water management/ irrigation, smart sensors, artificial intelligence (AI) and the Internet of Things (IoT). Even though this technology increases agricultural efficiency, it has, however, several limitations, such as high initial cost and the need for technical expertise. Growing crops in smart hydroponic environments is an avenue that still needs to be explored by researchers as there is minimal information available on this area but has a huge potential to address global food security [99].

7. Challenges and limitations linked to hydroponic systems

Even though hydroponic cultivation method has emerged as a promising system to produce food sustainably regardless of the season, it has its drawbacks. The initial setup cost and technical expertise are major challenges in hydroponic cultivation. One has to be financially ready when starting a hydroponic operation, as there is a significant amount of cost that is associated with the startup of the operation [64]. One of the basic requirements in hydroponic production is the expertise for the management of nutrients, lights and pests [1]. Even during the installation phase of the controlled environment, technical expertise is required, meaning that highly skilled personnel need to be hired to assemble the building materials. On the other side, energy is vital for the proper functioning of the hydroponic system. To optimize the production, pumps for continuous water circulation, filters and other energy-consuming components are necessary. In the case of a smart hydroponic system, more energy is required. To address energy-related issues, the installation of solar panels is recommended, even though this comes at a higher cost. Nutrient management is another key issue, as nutrient availability is dependent on water pH. It is not easy to maintain neutral soil pH in some other types of hydroponics.

8. Conclusions

In most countries, agricultural production is threatened by a rapid increase in human population as the per capita land availability decreases leaving limited arable space for agricultural activities. Most farmers rely on open-field agriculture, which has several limitations, such as labour and water use for crop production. On the other hand, human activities like the burning of fossil fuel are the major causes of environmental issues, which are a serious threat to food production systems. In this review, the alternative cost-effective method of growing crops for global sustainable food production is extensively discussed. Most farmers are transitioning from the conventional or soil-based methods to hydroponic systems. The review has shown an increase in the number of crops that grow successfully in hydroponic systems, surprisingly latter systems also allow the growth of medicinal and ornamental plants. The review further compared the response of food crops to both conventional and hydroponic systems and revealed that the hydroponic system reduces time to maturity and improves yield and size of crops, on top of that production in this system is all year round. The use of technology in hydroponics is also highlighted as modern agriculture advocates for innovation to produce maximum yield. Based on this, farmers can safely transition to this growing system to guarantee more yield throughout the year. The effect of this system on fruit quality is, however, not clear and still needs further investigation. It must be noted that there are limitations attached to the hydroponic system but they are minimal as compared to soil-based agriculture. Lastly, there is a need to find innovative and environmentally friendly methods that have a potential to partially reduce the use of chemical-based fertilizers in hydroponics for a sustainable production of crops.

Conflict of interest

The authors declare no conflict of interest.

Author details

Bonga Lewis Ngcobo*, Nokuzola Phungula, Philile Ngcobo and Zandile Maninjwa
Faculty of Applied Sciences, Department of Horticulture, Durban University of Technology, Durban, South Africa

*Address all correspondence to: bongan@dut.ac.za

References

[1] Gashgari R et al. Comparison between growing plants in hydroponic system and soil based system. In: Proceedings of the 4th World Congress on Mechanical, Chemical, and Material Engineering. Spain: ICMIE Madrid; 2018

[2] Zeifman L. et al. A World of 8 Billion. UN DESA: United Nations Department of Economic and Social Affairs; 2022

[3] Tudi M et al. Agriculture development, pesticide application and its impact on the environment. International Journal of Environmental Research and Public Health. 2021;**18**(3):1112

[4] Sharma N et al. Hydroponics as an advanced technique for vegetable production: An overview. Journal of Soil and Water Conservation. 2018;**17**(4):364-371

[5] Atherton HR, Li P. Hydroponic cultivation of medicinal plants—Plant organs and hydroponic systems: Techniques and trends. Horticulturae. 2023;**9**(3):349

[6] Park Y, Williams KA. Organic hydroponics: A review. Scientia Horticulturae. 2024;**324**:112604

[7] Zhu Z et al. Integrated hydroponics systems with anaerobic supernatant and aquaculture effluent in desert regions: Nutrient recovery and benefit analysis. Science of the Total Environment. 2023;**904**:166867

[8] Rajendran S et al. Hydroponics: Exploring innovative sustainable technologies and applications across crop production, with emphasis on potato mini-tuber cultivation. Heliyon. 2024;**10**(5):e26823

[9] Khan S, Purohit A, Vadsaria N. Hydroponics: Current and future state of the art in farming. Journal of Plant Nutrition. 2020;**44**(10):1515-1538

[10] Mariyappillai A, Arumugam G, Raghavendran V. The techniques of hydroponic system. Acta Scientific Agriculture. 2020;**4**(7):79-84

[11] Groenveld T et al. Optimization of nitrogen use efficiency by means of fertigation management in an integrated aquaculture-agriculture system. Journal of Cleaner Production. 2019;**212**:401-408

[12] Goddek S et al. Aquaponics Food Production Systems: Combined Aquaculture and Hydroponic Production Technologies for the Future. Switzerland: Springer Nature; 2019

[13] Goh YS et al. A meta-analysis: Food production and vegetable crop yields of hydroponics. Scientia Horticulturae. 2023;**321**:112339

[14] Woznicki T et al. Growing potatoes (*Solanum tuberosum* L.) hydroponically in wood fiber—a preliminary case-study report. Agronomy. 2021;**11**(7):1369

[15] Bradley P, Marulanda C. Simplified hydroponics to reduce global hunger. In: World Congress on Soilless Culture: Agriculture in the Coming Millennium. Acta Horticulturae; 2000;**554**:289-296

[16] Gebreegziher WG. Soilless culture technology to transform vegetable farming, reduce land pressure and degradation in drylands. Cogent Food & Agriculture. 2023;**9**(2):2265106

[17] Klerkx L, Rose D. Dealing with the game-changing technologies of

agriculture 4.0: How do we manage diversity and responsibility in food system transition pathways? Global. Global Food Security. 2020;**24**:100347

[18] Macwan J et al. Review on soilless method of cultivation: Hydroponics. International Journal of Scientific Research. 2020;**11**:37122-37127

[19] Ramos C et al. Experimental NFT hydroponics system with lower energy consumption. In: 2019 5th Experiment International Conference (Exp at'19). Funchal, Portugal: IEEE; 2019

[20] Gillani SA et al. Comparison of energy-use efficiency for lettuce plantation under nutrient film technique and deep-water culture hydroponic systems. Procedia Computer Science. 2023;**217**:11-19

[21] AlShrouf A. Hydroponics, aeroponic and aquaponic as compared with conventional farming. American Scientific Research Journal for Engineering,Technology,& Sciences. 2017;**27**(1):247-255

[22] Ferrarezi RS, Testezlaf R. Performance of wick irrigation system using self-compensating troughs with substrates for lettuce production. Journal of Plant Nutrition. 2016;**39**(1):147-161

[23] Lee S, Lee J. Beneficial bacteria and fungi in hydroponic systems: Types and characteristics of hydroponic food production methods. Scientia Horticulturae. 2015;**195**:206-215

[24] Dholwani SJ et al. Introduction of hydroponic system and it's methods. International Journal for Research Trends and Innovation. 2018;**3**(3):69-73

[25] Urbina MA et al. Adsorption of polyethylene microbeads and physiological effects on hydroponic maize. Science of the Total Environment. 2020;**741**:140216

[26] Kumar K et al. Evaluation of spinach and lettuce production (growth, morphological characteristics, biomass yield, and nutritional quality) in NFT hydroponic system in greenhouse, room, and open environment at Leh, India. European Chemical Bulletin. 2023;**12**(5):5421-5435

[27] Mohammed SB, Sookoo R. Nutrient film technique for commercial production. Agricultural Science Research Journal. 2016;**6**(11):269-274

[28] Hamza A et al. Using deep water culture as one of the important hydroponic systems for saving water, mineral fertilizers and improving the productivity of lettuce crop. International Journal of Health Sciences. 2022;**6**(59):2311-2331

[29] Majid M et al. Evaluation of hydroponic systems for the cultivation of lettuce (*Lactuca sativa* L., var. Longifolia) and comparison with protected soil-based cultivation. Agricultural Water Management. 2021;**245**:106572

[30] Nursyahid A et al. Analysis of deep water culture (DWC) hydroponic nutrient solution level control systems. In: IOP Conference Series: Materials Science and Engineering. Semarang, Indonesia: IOP Publishing; 2021

[31] Li Q et al. Growth responses and root characteristics of lettuce grown in aeroponics, hydroponics, and substrate culture. Horticulturae. 2018;**4**(4):35

[32] Chandra S et al. Assessment of total phenolic and flavonoid content, antioxidant properties, and yield of aeroponically and conventionally grown leafy vegetables and fruit crops: A comparative study. Evidence-based

Complementary and Alternative Medicine. 2014;**2014**(1):253875

[33] Çalışkan ME et al. Comparison of aeroponics and conventional potato mini tuber production systems at different plant densities. Potato Research. 2021;**64**:41-53

[34] George P, George N. Hydroponics-(soilless cultivation of plants) for biodiversity conservation. International Journal of modern Trends in Engineering and Science. 2016;**3**(06):97-104

[35] Kannan M et al. Hydroponic farming–A state of art for the future agriculture. Materials Today: Proceedings. 2022;**68**:2163-2166

[36] Birkby J. Vertical farming. ATTRA Sustainable Agriculture. 2016;**2**(1):1-12

[37] Avgoustaki DD, Xydis G. How energy innovation in indoor vertical farming can improve food security, sustainability, and food safety? In: Advances in Food Security and Sustainability. Amsterdam, The Netherlands: Elsevier; 2020. pp. 1-51

[38] Kaur G, Chawla P. All about vertical farming: A review. Turkish Journal of Computer And Mathematics Education. 2021;**12**(2):1-14

[39] Beacham AM, Vickers LH, Monaghan JM. Vertical farming: a summary of approaches to growing skywards. The Journal of Horticultural Science and Biotechnology. 2019;**94**(3):277-283

[40] Banerjee C, Adenaeuer L. Up, up and away! The economics of vertical farming. Journal of Agricultural Studies. 2014;**2**(1):40-60

[41] Sardare MDaASV. A review on plant without soil-hydroponics. International Journal of Research in Engineering and Technology. 2013;**2**:299-304

[42] Martin MaM E. Environmental assessment of an urban vertical hydroponic farming system in Sweden. Sustainability. 2019;**11**:1-14

[43] Singh SaS B. Hydroponics–A technique for cultivation of vegetables and medicinal plants. In: Proceedings of Proceedings of 4th Global Conference on Horticulture for Food, Nutrition and Livelihood Options. Bhubaneshwar, Odisha, India. 2012

[44] Keller C, Davidian JC, Pokrovsky OS, Bovet N, Chaurand P, Meunier JD. Effect of silicon on wheat seedlings (*Triticum turgidum* L.) grown in hydroponics and exposed to 0 to 30 μM Cu. Planta. 2015;**241**:847-860

[45] Halbert-Howard A, Häfner F, Karlowsky S, Schwarz D, Krause A. Evaluating recycling fertilizers for tomato cultivation in hydroponics, and their impact on greenhouse gas emissions. Environmental Science and Pollution Research. 2021;**28**:59284-59303

[46] Sajiv G, Anburani A, Venkatakrishnan D. Study on the growth of eggplant (*Solanum melongena* L.) under hydroponics with modified Hoagland solution. Journal of Current Research in Food Science. 2023;**4**:04-06

[47] Lennard W, Ward J. A comparison of plant growth rates between an NFT hydroponic system and an NFT aquaponic system. Horticulturae. 2019;**5**(2):27

[48] Gurung CT, Bhandari JB, Gurung A. Evaluation of hydroponic cultivation techniques as a supplement to conventional methods of farming.

Journal of Agriculture and Technology. 2019;**6**(1):57-64

[49] Walters KJ, Currey CJ. Hydroponic greenhouse basil production: Comparing systems and cultivars. HortTechnology. 2015;**25**(5):645-650

[50] Ai P et al. Comparison of chrysanthemum flowers grown under hydroponic and soil-based systems: Yield and transcriptome analysis. BMC Plant Biology. 2021;**21**:1-17

[51] Sharma D et al. Hydroponic cultivation enhances the morpho-physiological traits and quality flower production in three cultivars (marigold scarlet red, marigold orange, and marigold yellow) of French marigold (*Tagetes patula* L.). Scientia Horticulturae. 2024;**327**:112803

[52] Shao M et al. Differential effects of high light duration on growth, nutritional quality, and oxidative stress of hydroponic lettuce under red and blue LED irradiation. Scientia Horticulturae. 2020;**268**:109366

[53] Kandeel M et al. Genome-wide identification of B3 DNA-binding superfamily members (ABI, HIS, ARF, RVL, REM) and their involvement in stress responses and development in Camelina sativa. Agronomy. 2023;**13**(3):648

[54] Bafort F et al. The agro-economic feasibility of growing the medicinal plant Euphorbia peplus in a modified vertical hydroponic shipping container. Horticulturae. 2022;**8**(3):256

[55] Bose S, Alam MK, Hossain AZ, Hossain AK, Mymensingh B, Rashid MH. Effects of salinity on seedling growth of four maize (*Zea mays* L.) cultivars under hydroponics. Journal of Agricultural. 2018;**6**:56-69

[56] Swain A, Chatterjee S, Vishwanath M. Hydroponics in vegetable crops: A review. The Pharma Innovation Journal. 2021;**10**(6):629-634

[57] Treftz C, Omaye ST. Comparison between hydroponic and soil systems for growing strawberries in a greenhouse. International Journal of Agricultural Extension. 2016;**3**(3):195-200

[58] Lei C, Engeseth NJ. Comparison of growth characteristics, functional qualities, and texture of hydroponically grown and soil-grown lettuce. Lwt. 2021;**150**:111931

[59] Verdoliva SG et al. Controlled comparisons between soil and hydroponic systems reveal increased water use efficiency and higher lycopene and β-carotene contents in hydroponically grown tomatoes. Scientia Horticulturae. 2021;**279**:109896

[60] Chandra S et al. Assessment of total phenolic and flavonoid content, antioxidant properties, and yield of aeroponically and conventionally grown leafy vegetables and fruit crops: a comparative study. Evidence-based Complementary and Alternative Medicine. 2014;**2014**:1-9

[61] Lazo RP, Gonzabay JQ. Economic analysis of hydroponic lettuce under floating root system in semi-arid climate. La Granja. 2020;**31**(1):118

[62] Purba JH, Parmila IP, Dadi W. Effect of soilless media (hydroponic) on growth and yield of two varieties of lettuce. Agricultural Science. 2021;**4**(2):154-165

[63] Gumisiriza MS et al. Soilless urban gardening as a post covid-19 food security salvage technology: A study on the physiognomic response of lettuce to hydroponics in Uganda. Scientific African. 2023;**20**:e01643

[64] Kumar S et al. Opportunities and constraints in hydroponic crop production systems: A review. Environment Conservation Journal. 2021;**22**(3):401-408

[65] Kratky B. Three non-circulating hydroponic methods for growing lettuce. In: International Symposium on Soilless Culture and Hydroponics 843. 2008

[66] Kratky B. Growing lettuce in non-aerated, non-circulated hydroponic systems. Journal of Vegetable Science. 2005;**11**(2):35-42

[67] Folorunso EA et al. The economic viability of commercial-scale hydroponics: Nigeria as a case study. Heliyon. 2023;**9**(8):18979

[68] Mishra T. Climate change and production of secondary metabolites in medicinal plants: a review. International Journal of Herbal Medicine. 2016;**4**(4):27-30

[69] Arora NK. Impact of climate change on agriculture production and its sustainable solutions. Environmental Sustainability. 2019;**2**(2):95-96

[70] Liu K et al. Silver lining to a climate crisis in multiple prospects for alleviating crop waterlogging under future climates. Nature Communications. 2023;**14**(1):765

[71] Crumpler K, Bernoux M. Climate change adaptation in the agriculture and land use sectors: a review of nationally determined contributions (NDCs) in Pacific Small Island developing states (SIDS). Managing Climate Change Adaptation in the Pacific Region. 2020:1-25

[72] Ruane AC, Rosenzweig C. Climate change impacts on agriculture. In: Agriculture & Food Systems to 2050: Global Trends, Challenges and Opportunities. Singapore: World Scientific; 2019. pp. 161-191

[73] Li G et al. Design and performance of a novel rice hydroponic biofilter in a pond-scale aquaponic recirculating system. Ecological Engineering. 2018;**125**:1-10

[74] Bakri H, Abou-Shleel S. Economic impacts of climatic changes on some vegetable crops in Egypt. World Journal of Agricultural Sciences. 2013;**9**:361-368

[75] De Wrachien D, Goli MB. Global warming effects on irrigation development and crop production: A world-wide view. Agricultural Sciences. 2015;**6**(07):734

[76] Chen B et al. Global land-water nexus: Agricultural land and freshwater use embodied in worldwide supply chains. Science of the Total Environment. 2018;**613**:931-943

[77] Putra PA, Yuliando H. Soilless culture system to support water use efficiency and product quality: A review. Agriculture and Agricultural Science Procedia. 2015;**3**:283-288

[78] Pomoni DI et al. A review of hydroponics and conventional agriculture based on energy and water consumption, environmental impact, and land use. Energies. 2023;**16**(4):1690

[79] Villarroel M, Rodriguez Alvariño JM, Duran Altisent JM. Aquaponics: Integrating fish feeding rates and ion waste production for strawberry hydroponics. Spanish Journal of Agricultural Research. 2011;**9**(2):537-545

[80] Girija P et al. Location and technology options to reduce environmental impacts from agriculture. Journal of Cleaner Production. 2014;**81**:130-136

[81] Chausali N, Saxena J. Conventional versus organic farming: Nutrient status. In: Meena VS, Meena SK, Rakshit A, Stanley J, Srinivasarao C, editors. Advances in Organic Farming. Bipin Tripathi Kumaon Institute of Technology. Almora, India: Woodhead Publishing; 2021

[82] Velazquez-Gonzalez RS, Garcia-Garcia AL, Ventura-Zapata E, Barceinas-Sanchez JDO, Sosa-Savedra JC. A review on hydroponics and the technologies associated for medium-and small-scale operations. Agriculture. 2022;**12**:1-21

[83] Kwon MJ et al. Waste nutrient solutions from full-scale open hydroponic cultivation: Dynamics of effluent quality and removal of nitrogen and phosphorus using a pilot-scale sequencing batch reactor. Journal of Environmental Management. 2021;**281**:111893

[84] Baddadi S et al. Autonomous greenhouse microclimate through hydroponic design and refurbished thermal energy by phase change material. Journal of Cleaner Production. 2019;**211**:360-379

[85] Barbosa GL et al. Comparison of land, water, and energy requirements of lettuce grown using hydroponic vs. conventional agricultural methods. International Journal of Environmental Research and Public Health. 2015;**12**:6879-6891

[86] Velazquez-Gonzalez RS et al. A review on hydroponics and the technologies associated for medium-and small-scale operations. Agriculture. 2022;**12**(5):646

[87] Nabi S et al. Hydroponics: Environmentally sustainable practice in the agricultural system. Pharma Innovation Journal. 2022;**11**:207-212

[88] Tomasi N et al. New 'solutions' for floating cultivation system of ready-to-eat salad: A review. Trends in Food Science & Technology. 2015;**46**(2):267-276

[89] Lopez-Galvez F et al. Monitoring generic Escherichia coli in reclaimed and surface water used in hydroponically cultivated greenhouse peppers and the influence of fertilizer solutions. Food Control. 2016;**67**:90-95

[90] Salgot M, Folch M. Wastewater treatment and water reuse. Current Opinion in Environmental Science & Health. 2018;**2**:64-74

[91] Naresh R et al. Role of hydroponics in improving water-use efficiency and food security. International Journal of Environment and Climate Change. 2024;**14**(2):608-633

[92] Naznin MT, Lefsrud M. An overview of LED lighting and spectral quality on plant photosynthesis. In: Light Emitting Diodes for Agriculture: Smart Lighting. Singapore: Springer; 2017. pp. 101-111

[93] Ngcobo BL, Bertling I. An overview of the recent developments in the postharvest application of light-emitting diodes (LEDs) in horticulture. New Advances in Postharvest Technology. 2023

[94] Ngcobo BL, Bertling I. An overview of the recent developments and current status on the preharvest application of LED technology in controlled environment agriculture. The Journal of Horticultural Science and Biotechnology. 2024:1-8

[95] Li J et al. Effect of LED spectrum on the quality and nitrogen metabolism of lettuce under recycled hydroponics. Frontiers in Plant Science. 2021;**12**:678197

[96] Talukder MR et al. Light-emitting diodes and exogenous amino acids application improve growth and yield of strawberry plants cultivated in recycled hydroponics. Scientia Horticulturae. 2018;**239**:93-103

[97] Bantis F et al. Current status and recent achievements in the field of horticulture with the use of light-emitting diodes (LEDs). Scientia Horticulturae. 2018;**235**:437-451

[98] Lakshmanan R et al. Automated smart hydroponics system using internet of things. International Journal of Electrical and Computer Engineering (IJECE). 2020;**10**(6):6389-6398

[99] Andrianto H, Ahmad F. Development of smart greenhouse system for hydroponic agriculture. In: 2020 International Conference on Information Technology Systems and Innovation. Bandung - Padang; 2020. pp. 335-340

Chapter 5

Hydroponic Farming: Innovative Solutions for Sustainable and Modern Cultivation Technique

Gamachis Korsa, Abate Ayele, Setegn Haile and Digafe Alemu

Abstract

As conventional soil-based farming face limitations due to diminishing arable land per capita, advanced agricultural technologies have emerged as a promising solutions. Among these, hydroponic farming – a soilless crop cultivation method – stands out as a leading innovation in vegetable production, offering a viable response to these pressing challenges. This chapter explores the world of hydroponic farming, highlighting the best and most sustainable practices associated with this modern cultivation technique. By reviewing peer-reviewed articles from reputable educational journals, the chapter categorizes the findings into four key areas: types of hydroponic farming systems, factors that affect their performance, substrate constituents, and potential applications in modern agriculture. The findings of the current review indicate that hydroponic farming is an effective tool for combating hunger and improving food safety, especially in developing countries with limited water resources. By implementing innovative techniques that enhance resource utilization, reduce health impact, and create a more sustainable for food production, hydroponics represents a significant advancement in agriculturral environment.

Keywords: cultivation, hydroponic, soil-less farming, substrate, sustainable

1. Introduction

The rapidly expanding global population is placing a significant strain on food security, a vital pillar of national prosperity and economic growth. As the world's population continues to swell, the availability and accessibility of nutritious food have become a pressing concern, directly impacting a country's overall well-being and economic stability. The escalating demand for food necessitates the use of more energy, water, and soil resources, which can have negative environmental consequences. The widespread adoption of intensive farming practices can lead to soil degradation, nutrient depletion, and reduced soil fertility, ultimately threatening the long-term productivity of the land [1, 2].

The global population has more than doubled since 1960 and is anticipated to reach 9.8 billion by 2050. Although between 2015 and 2018 the growth rate remained relatively stable at 10.6 and 10.8% respectively, the number of people suffering from undernourishment increased by 4.6%, rising from 785.4 million in 2015 to 821.6 million in 2018. This population increase intensifies the pressing issue of food security, which is a crucial element of sustainability. Consequently, investing in agriculture is critical to ensure food security and combat global hunger and poverty [3]. Conventional farming practices face numerous challenges, such as inefficient irrigation water usage, the requirement for large tracts of land, and dependence on chemical fertilizers. Consequently, soilless agriculture has emerged as a promising alternative that addresses these issues. In addition to conserving and rehabilitating agricultural land, soilless farming – particularly when employing closed-loop systems – offers numerous benefits: it recycles a predetermined amount of water, reduces irrigation water consumption by 85–90%, can be utilized in locations unsuitable for traditional farming, generates minimal environmental pollution, and generally results in higher yields compared to conventional methods [4].

.As the availability of arable land declines and environmental factors become more challenging, farmers are increasingly turning to innovative soilless farming methods to cultivate crops, particularly vegetables. Soilless cultivation involves growing plants in mediums other than soil, using irrigation water to supply essential nutrients. This approach offers a range of benefits, including more sustainable agricultural practices, improved soil conservation, and advanced industrial techniques. In hydroponics, essential nutrients are delivered through mediums such as rock wool, hydroton, and perlite for various crops, including leaf lettuce, tomatoes, peppers, and cucumbers. On the other hand, aeroponics is a subset of hydroponics where plants are suspended in the air, and their roots are periodically misted with a nutrient solution from an automated sprinkler system connected to a central nutrient reservoir, ensuring efficient and accurate delivery. This method is effective for growing a wide range of vegetables and fruits, including beets, broccoli, cabbage, carrots, potatoes, tomatoes, and watermelon [5, 6].

The term "hydroponics" originates from the combination of the Greek words "hydro," meaning water, and "ponos," meaning labor. Hydroponics involves growing plants in nutrient-rich water, often without the need for a solid medium like soil. Unlike traditional farming methods, hydroponic plants do not require a solid foundation, instead receiving a balanced mix of nutrients dissolved in water. One of the key advantages of hydroponics is the reduced labor costs, as the process typically takes place in controlled environments with automated irrigation and fertilization systems [7].

In densely populated urban areas, hydroponic farming is an ideal solution due to limited land availability. By using rooftop and vertical hydroponic systems, farmers can cultivate a variety of vegetables, reducing the distance food travels and increasing local food security. This innovative approach yields 1000 times more produce than conventional methods, making it an attractive solution for compact cities. In addition to issues like low soil fertility in various arable regions, limited natural soil fertility replenishment by microorganisms resulting from ongoing cultivation, frequent droughts, unpredictable climate and weather conditions, rising temperatures, river pollution, ineffective water management, and significant water waste, as well as decreasing groundwater levels, conventional soil-based agriculture faces substantial threats to food production. If these challenges persist, it may soon become unfeasible to feed the entire population [8, 9]. The evolution of hydroponics has been a significant development in modern agriculture, holding immense potential to transform the way we grow crops. From its historical roots to its current widespread adoption, hydroponics has evolved as a sustainable and innovative approach to address the pressing issues of food security and environmental sustainability [10].

This innovative approach enables the production of a large quantity of crops, vegetables, and plants in a controlled environment. One of the significant benefits of hydroponics is the superior quality of the harvested products, which boasts better taste, nutritional value, and overall yield compared to soil-based cultivation. Additionally, hydroponics offers several advantages, including reduced costs, decreased risk of disease, and eco-friendliness [11]. Hydroponics is also a contemporary agricultural method that eliminates the need for soil in crop cultivation, earning it the label of soilless farming. This approach is an effective alternative for producing nutritious fruits and vegetables, as it reduces exposure to soil-borne pathogens by not utilizing soil for growth. Hydroponics aligns with the 2030 Sustainable Development Goals (SDGs) zero hunger agenda, enabling individuals to grow their food at home. This initiative has the potential to empower young people and women, thereby contributing to the country's economic growth. The use of the root dipping technique in hydroponic farming has been found to promote exceptional growth and high yields in jute mallow production [12].

Compared to conventional farming techniques, hydroponics has several advantages, including reduced water consumption – typically using 90% less water. The popularity of hydroponics stems from its cost-effectiveness and health benefits for consumers. By growing plants without soil, hydroponics eliminates the risks of pests and diseases, eliminating the need for expensive treatments and preservation methods. Soilless cultivation is increasingly adopted in protected agriculture to enhance the management of the growing environment and minimize unpredictability in soil water and nutrient conditions. Recently, the method of soilless culture has shifted from an open to a closed-loop system, which is recognized for improved water use efficiency while preserving yield quality [13, 14]. The integration of next-generation technologies in soilless vegetable production is poised to revolutionize yield, efficiency, and sustainability. Artificial intelligence (AI)-powered monitoring systems and precision farming methods enable the precise tracking of crucial factors such as nutrient levels and pH in soilless vegetable production, ultimately informing the successful implementation of cutting-edge technologies that can optimize crop growth and reduce environmental impact [15].

Despite the success of hydroponics on a large scale, implementing this method on a small scale, particularly in urban and suburban settings, poses unique challenges. In rural communities, where access to suitable technologies is limited, the adoption of hydroponics can be even more daunting. However, recent advancements in technologies such as the Internet of Things (IoT) and Industry 4.0 have paved the way for precision agriculture on a small scale. This approach enables farmers to monitor and control key variables like pH, electrical conductivity, and temperature with precision, resulting in increased yields and reduced resource consumption [16].

New approaches to soilless production, such as hydroponic farming, hold great potential to solve agricultural issues related to sustainability, productivity, and efficient use of resources. It describes the start expenses, ongoing costs, and potential revenue of hydroponic systems with traditional soil-based farming to assess the economic viability of these systems through a comprehensive cost-benefit analysis. This technology allows for sustainable agricultural methods and assists in achieving the growing requirements of the world's food systems [17].

The chapter conveys that hydroponic farming is a viable and sustainable option for modern agriculture. By adopting best practices, such as efficient nutrient management, water conservation, environmental control, and responsible resource use, hydroponics can significantly contribute to sustainable food production. This method not only supports the demands of a growing population but also addresses environmental concerns, making it an essential aspect of the future of farming. Through innovation and adherence to sustainable practices, hydroponic systems can lead to increased food security and healthier ecosystems. Therefore, this chapter focused on hydroponic farming – best and sustainable practices for modern cultivation techniques.

2. Overview of hydroponic farming system for modern cultivation techniques

The requirement for innovative agricultural techniques to be developed continues to be continually in demand, particularly because of the numerous difficulties that traditional farming encounters. Hydroponics is a unique advancement technology that has solved these issues. High-yielding hydroponic systems can be placed strategically close to population centers by farmers, which will reduce shipping time and fertilizer loss [18]. Soilless agriculture can thrive in various locations, such as balconies, rooftops, greenhouses, and areas that are not suitable for traditional farming. In this method, fully soluble nutrients are dissolved in water and delivered to the plants using various irrigation technologies, such as injection systems. The plants are supported using above-ground structures like benches, beds, and troughs were well-suited for cultivating fruiting and leafy vegetables, strawberries, and ornamental plants [19].

In soilless cultivation, crops rely on a nutrient-rich solution for growth, as they do not come into contact with soil and precipitation. To ensure healthy plant development and optimal yields, this solution must be supplied in excess, which results in a significant amount of excess water and nutrients being removed from the root systems [20]. Hydroponics reduces water use by up to 90% and lowers the risk of nutrient runoff and soil erosion by doing away with the requirement for soil. Compared to conventional farming methods, hydroponic systems' controlled environment allows for exact monitoring of temperature, light, and nutrient levels, which enhances plant growth and increases yields. Hydroponics enables continuous cultivation throughout the year, making it possible to grow crops in areas with challenging climates or limited farmland. The environmentally friendly aspects of hydroponics go beyond saving water and optimizing land usage. By carefully regulating nutrient solutions in hydroponic systems, growers can apply fertilizers more precisely, reducing waste and minimizing the release of harmful chemicals into the ecosystem [21].

2.1 Types of hydroponics farming system in modern farming

The acceptance of hydroponics technology as a sustainable urban farming solution is gaining traction, attracting a diverse range of stakeholders including technologists, non-farmers, retailers, restaurants, and consumers. While hydroponics offers an innovative approach to food production, its environmental impact remains unclear. The global hydroponics market is projected to reach $25.1 billion by 2027, comprising various entities such as food manufacturers, farm producers, hydroponic system suppliers, distributors of hydroponics components, and other key players [22].

The process of growing plants without soil utilizing mineral nutrient solutions and/or soilless substrates like sawdust, gravel, sand, pumice, perlite, vermiculite, rock wool, peat moss, or coco coir is commonly referred to as hydroponics or hydroponic systems. A variety of hydroponic systems may be obtained, as indicated in **Table 1**, depending on how the fertilizer solution is applied. They include deep water culture (DWC), nutrient film technique (NFT), wick, drip, ebb-flow, and aeroponic configurations [11, 43, 44].

2.2 Other modern agricultural activities by hydroponic system

The occurrence of hydroponics systems, aeroponics, and aquaponics has revolutionized the agricultural sector by offering innovative alternatives to traditional soil-based farming. These cutting-edge methods have proven to be more efficient, productive, and environmentally friendly. By utilizing resources more effectively, increasing crop yields, and minimizing environmental impact, these modern cultivation techniques have transformed the way we grow our food [45]. Recently, the development of modern agricultural technologies has revolutionized farming practices, allowing for increased efficiency and novel production methods. These cutting-edge systems not only optimize land use but also conserve water and provide a controlled environment for optimal crop growth [46].

The SDGs, specifically Goal 2 (Zero Hunger) and Goal 12 (Responsible Consumption and Production), highlight the critical role of sustainable agriculture in achieving global well-being. As the world grapples with rapid population growth and urbanization, the need for efficient food production methods that conserve vital resources like water and energy has become more pressing than ever [47]. Hydroponics is a method of growing plants in a controlled nutrient-rich solution, often supplemented with an inert medium like gravel, vermiculite, or rock wool to provide structural support. This approach allows for the development of more industrialized and automated systems that can boost productivity while promoting ecological balance and sustainability. By integrating facilities for environmental protection and improvement, hydroponics can contribute to socio-economic growth and development [48–50].

The demanding need for increased food production, combined with limited natural resources and arable land, and recent restrictions on energy consumption necessitate an urgent solution in agricultural practices. To assess the environmental impact of conventional agriculture versus hydroponics, we evaluated factors such as soil degradation, crop-soil contamination, and greenhouse gas emissions. In terms of resource consumption, it compared conventional agriculture to hydroponics based on water usage rates per kilogram of produce (L/kg), energy consumption rates per kilogram of produce (kWh), and the total energy required (kW) [51].

By reducing reliance on arable land and pesticides, hydroponics can thrive in areas with poor soil quality, minimizing the impact of extreme weather events. As the world's population grows at an unprecedented rate, hydroponics offers a potential solution for growing vegetables in urban areas. With controlled conditions, nutrient-rich substrates, and solid support systems, hydroponics is poised to expand globally, even in challenging agro-climatic zones [52].

2.2.1 Aeroponics system

The term "aeroponic" originates from the Greek words "aero" meaning "air" and "ponos" meaning "work". An aeroponic system is a self-contained or partially closed

Types of system	Characterics of hydroponic systems	Reference
Drip system	This system used for globally soil-based agriculture, measure drops/the evaporation of water reduce, and therefore, the plant roots are always kept moist without overwatering or runoff as in the case of conventional agriculture.	[22]
	These drip system methods offer numerous benefits, including shorter crop growing durations compared to conventional farming, the ability to produce crops year-round, reduced occurrence of diseases and pests, and the removal of the need for weeding, spraying, and watering.	[23]
Ebb and flow system	The most preferred for beginners; even experts opt for this method and does not require any machinery and the production of crops.	[24]
	Ebb denotes the period of water recession and drainage, using a pump, submerging the roots in a nutrient solution for a brief period (5 to 10 minutes), followed by deactivation to allow solution evacuation.	[25]
	In this system, the supply water is pumped from the water and nutrient management storage tank to each bench or group of benches, filling to a depth of 1–2 cm within 5 min and draining within 10 min for a total water cycle per bay of 15 min.	[26]
Deep water culture	It is commonly known as the deep flow technique, floating raft technology. Alternatively, sufficient quantities of hydrogen peroxide (H_2O_2) can be used.	[27]
	By using these methods, cultivation of crops can be reduced by 75%, and also the requirement of water for irrigation can be lowered by 90% with negligible environmental resource.	[28]
	The plants are suspended above the water using a support system such as net cups, and their roots are submerged in the water. One of the main advantages of the DWC system is its simplicity.	[29]
Nutrient film technique (NFT)	This allows the roots to absorb the necessary nutrients and oxygen, promoting healthy plant growth. In NFT systems, the roots of the plants are suspended in the nutrient solution, and the water is continuously flowing over the roots.	[30]
	Hydroponic systems, particularly nutrient film technique (NFT), are highly effective due to their soilless nature and efficient nutrient and water circulation.	[31]
Wick system	The wick system acts as a conduit, delivering nutrient solution to the roots of potted plants.	[32]
	Wick system so that the production costs of hydroponic cultivation can be cheaper and more affordable.	[33]
Aeroponic system	In this system, roots are held in a soilless growing medium like Coco coir on the grow tray. On these roots, nutrient-rich solution is being sprayed by using specially designed misting devices.	[34]
	In aeroponic systems, plants are grown on styrofoam plates that are secured to a frame that can be either level or inclined at a precise angle of 45 to 60 degrees, commonly referred to as an A-frame design.	[35]
	In aeroponics techniques, plants receive higher levels of oxygen, which helps enhance their food-producing activity and makes them less susceptible to pests and diseases.	[36]
	Aeroponics is very suitable for studying water stress and root morphology due to its ability to precisely adjust the moisture content in the root zone and the amount of water provided	[37, 38]
Aquaponics system	A wide range of different agencies have jurisdiction over water, animal health, environmental protection and food safety, and their regulations are in some cases contradictory or are ill-suited for complex integrated aquaponics system.	[39]

Types of system	Characterics of hydroponic systems	Reference
	The use of saline water in aquaponics is a relatively new development, and as with most new developments, the terms used to describe the range/hierarchy of types need to be established on a firm footing.	[40]
	The most pressing issues are whether aquaponics can become acceptable at the policy level. An emerging food production technology which has the ability to condense and compress production into spaces and places that would not normally be used for growing food by aquaponics system.	[41, 42]

DWC: deep water culture; NFT: nutrient film technique.

Table 1.
Types of hydroponic systems and their characterics in modern techniques.

system where plants are grown without the need for soil or an aggregate medium, with their roots suspended in a deep air or mist environment. This innovative method allows plants to grow rapidly, as the fine mist also delivers oxygen directly to the roots. As a result, the plants thrive in this unique environment, with the reservoir constantly replenished with fresh nutrients [53].

A novel approach to hydroponics has been developed by combining multiple irrigation systems, including drip irrigation, aeroponics, wick systems, and deep water culture systems, within a closed-loop system. This integrated approach offers significant benefits, including reduced water and nutrient loss to the environment. Compared to traditional open hydroponic systems, the closed system demonstrates remarkable improvements in water productivity, reaching up to 96%, and nutrient utilization, reaching up to 97% [54].

An aeroponic system utilizes a novel method of providing plants with essential nutrients. Instead of traditional soil, plants are cultivated in a nutrient-rich mist that is atomized by a nozzle and sprayed at regular intervals. This innovative approach allows for year-round production of fresh and nutritious fruits and vegetables, both indoors and on rooftops, using artificial light sources. The system consists of a growing chamber, net pots, an irrigation system, a motor, a digital timer, and a nutrient tank. This cutting-edge technique offers a sustainable and efficient way to grow crops in controlled environments [33, 55, 56].

With the growing challenges of food security and environmental degradation, there is an increasing need for sustainable agricultural practices. Soilless cultivation techniques offer a promising solution by minimizing land and water use while maximizing crop yield and quality. The world's population is projected to rise by nearly a billion people over the next 12 years, reaching 8.1 billion in 2025 and 9.6 billion by 2050. This rapid growth will lead to an increased demand for food production, placing a significant strain on the global agricultural system. Aeroponics offers a more efficient and controlled method of growing crops, addressing the limitations of traditional field farming practices [57].

2.2.2 Aquaponics system

The second SDG to achieve zero hunger by 2030 is under threat if current agricultural practices persist. Conventional farming methods are often inefficient and compete for space with a growing global population. In contrast, urban agriculture offers a more sustainable and productive solution by utilizing soilless or alternative growing mediums in urban areas. One such method gaining attention is aquaponics, which has evolved into decoupled or multi-loop systems to address the limitations of conventional approaches [58].

Urban aquaponics farms, which combine aquaculture with hydroponic vegetable production, offer a revolutionary and sustainable approach to food production in urban areas with limited agricultural land and water resources. However, the equipment employed in these systems, such as light-emitting diodes (LED) grow lights, air conditioning units, osmotic systems, fertilization pumps, and circulation pumps, are significant energy consumers. [59]. Aquaponics, a novel food production system, has the potential to revolutionize the way we grow food by combining the cultivation of aquatic animals with crops in a controlled environment. This innovative approach can significantly reduce water consumption and minimize waste generation. However, as a relatively new technology, aquaponics faces distinct regulatory and technical hurdles that are not encountered by traditional farming methods [60].

According to the National Aeronautics and Space Administration (NASA), aeroponics is a method of cultivating plants in the air without using soil or medium, enabling efficient and rapid food production. This innovative approach allows for year-round cultivation and harvesting without interruptions, as well as reduced risks of contamination from soil, pesticides, and residues. Additionally, aeroponic systems significantly minimize water usage by 98%, and fertilizer usage by 60%, and eliminate the need for pesticides. Moreover, studies have shown that plants grown through aeroponics tend to absorb more minerals and vitamins, resulting in healthier and potentially more nutritious produce [61].

2.3 Factors that affect hydroponic system performance

Plant factories require hydroponic systems, like the deep flow technique, nutrient film techniques, or aeroponic systems. Measurements of temperature, dissolved oxygen content, pH, electrical conductivity (EC), and pH are vital for proper water and fertilizer management in a hydroponic system. Real-time measurements of all nutrients are essential because changes in ion concentrations in nutrient solutions generate nutrient imbalances in closed hydroponic systems [62, 63] which are depicted in **Table 2**.

In hydroponics, the success of plant growth hinges on the meticulous management of the nutrient solution. Continuous monitoring and control of the nutrient levels in the tank are crucial to ensure optimal conditions for plant growth at all times. The implementation of an efficient automation algorithm is vital for a thriving hydroponic farm, as manual maintenance of farm parameters can be time-consuming and prone to human error, which can ultimately impact crop yields and quality [81]. It is essential to consistently manage and maintain the humidity levels in a growth chamber according to the requirements of the plants. The aeroponics system offers an optimal environment for oxygenation, enabling plant roots to develop in the air with abundant oxygen available [56].

Conventional agriculture often applies water excessively across extensive fields, leading to significant losses from runoff, evaporation, or poor absorption. In contrast, hydroponics takes a more sustainable approach to water usage. Hydroponic systems utilize a closed-loop method, where nutrient-enriched water is supplied directly to the roots of plants. Any water that is not absorbed is recycled within the system, reducing waste. As a result, hydroponics can use up to 90% less water than conventional soil-based farming methods. Therefore, hydroponics not only conserves water

Factors	Affects of hydroponic system	Reference
pH level	Hydroponic system has shown that the optimal the pH range for most plants is slightly acidic to neutral, with a range of approximately 5.5 to 6.5.	[64]
	A new nutrient solution management strategy, which examined pH of 4.0, 4.5, 5.0, and 5.5 with or without micronutrient adjustments for growing two cultivars of basil plants Dolce Fresca and Nufar in a greenhouse hydroponic deep-water culture (DWC) system.	[65]
	Plants grow best in a neutral pH nutrient media, and those grown hydroponically at pH levels of 5, 8, and 9 have shown drastic differences in their growth patterns, which can ultimately lead to the decline.	[66]
Electrical Conductivity	In hydroponics, low values of EC indicate a scarcity of nutrients in the form of ions; on the other hand, too-high values may lead to salt stress in the plant.	[67]
	The electrical conductivity (EC) has a significant influence on the health and vigor of crops in hydroponics.	[68]
	The electricity used to drive air conditioning systems to regulate the temperature and humidity of the greenhouse environment.	[69]
	This electricity was used to exploit and dispense water from the well in the irrigation strip and to ventilate the environment inside the greenhouse.	[70]
Temperature	It influences not only on the initial growing stage but also on harvesting period.	[71]
	The performance of this hydroponic greenhouse was evaluated without conventional heating methods, and results indicated that the hydroponic greenhouse/controlled provides better environmental conditions than conventional greenhouses.	[72]
	During the day, temperatures inside the greenhouse reached above 18°C and the temperature difference between outside and inside the greenhouse was 6°C.	[73]
	Maintaining the appropriate temperature for specific seeds during germination in a hydroponic system is crucial to monitoring.	[74]
Sun light	The combinations of red and blue wavelengths with high photon efficiency but that green and white light containing substantial amounts of green wavelengths has a positive physiological impact on plants.	[75]
	Light intensity significantly influences plant growth in hydroponic green fodder systems, yet research exploring the growth dynamics and nutrient accumulation in hydroponically grown barley under various light conditions has been limited.	[76]
	Root vegetables, such as carrots and potatoes, produced a lower demand for blue light but still required some development	[77]
	LED, high-intensity discharge (HID), or fluorescent lights are the most commonly used light sources for hydroponics.	[29]
Nutrients	A balanced nutrient solution is required in hydroponic systems, which should contain essential macronutrients such as nitrogen, phosphorus, and potassium, along with micronutrients like iron, zinc, and manganese, in appropriate concentrations.	[64]
	Fertilizers provide essential nutrients for plant growth, but excessive use can harm the environment. Smart farming technologies, such as the normalized difference vegetation index (NDVI), can help estimate the amount of fertilizer plants need without causing harm.	[78]

Factors	Affects of hydroponic system	Reference
	In hydroponics, all essential nutrients are provided to the plant via the nutrient solution, except for carbon, hydrogen, and oxygen, which are air-borne.	[16]
Water and Moisture	The seedling tool of hydroponic system is designed to monitor moisture levels: when the moisture falls below 60%, the pump activates, while it deactivates when the moisture rises above 60%	[79]
	Achieving the right moisture balance in the growing medium is essential for establishing seedlings successfully. It is advisable to moisten the selected medium prior to sowing the seeds, but care should be taken not to over-saturate it. For effective germination, keeping a steady moisture level is key for the seeds.	[80]
	Hydroponic systems used to conservation of water: it requires just 2–3 liters of water to produce 1 kg of lush green fodder, as compared to 60–80 liters to conventional system of fodder production.	[11]

DWC: deep-water culture, EC: electrical conductivity, LED: light-emitting diodes, NDVI: normalized difference vegetation index.

Table 2.
Factors that effective hydroponic system and their characterics in modern techniques.

but also optimizes its utilization, making sure every drop contributes to plant growth [82].

In the research conducted by Baddadi et al. [73], notable advancements were made in creating and constructing a microclimate within a greenhouse utilizing a hydroponic system along with regenerated thermal energy. The evaluation of this hydroponic greenhouse's performance was carried out without the use of traditional heating methods, with findings indicating that it maintained superior environmental conditions compared to standard greenhouses. Daytime temperatures within the greenhouse exceeded 18°C, and the temperature difference between the interior and exterior was recorded at 6°C. Relative humidity levels varied between 20 and 35% during the daytime and increased to 70–85% at night. The study also explored how environmental variables impact the energy balance in strawberry cultivation within the greenhouse, experimentally examining factors such as relative humidity, sunlight levels, and ambient temperature.

Zamora-Izquierdo et al. [83] created an advanced automated crop management system, which was evaluated in a greenhouse setting for tomato cultivation with a recirculation setup. This system comprised four main components: a nutrient solution preparation unit, an irrigation unit, a disinfection unit for drainage water (DW), and a purification unit for tap water. Within the nutrient solution preparation unit, a combination of nutrients, disinfected drainage water, tap water, and reverse osmosis-treated water was mixed. The nutrient solution's composition was automatically regulated based on electrical conductivity (EC), pH, and temperature, utilizing a specialized fertilization control module.

De Lucas Leal et al. [84] studied that a comparison was made between traditional greenhouses and hydroponic systems for spinach cultivation utilizing saline water for irrigation. The research determined that the implementation of saline water in agriculture necessitates the use of compatible cultivation methods. The study examined three types of cultivation: covered soil, uncovered soil, and a hydroponic system, using brackish water with electrical conductivity (EC) levels of 0.8, 1.5, 3.0, 4.5, 6.0, and 7.5 dS/m. Observations were made 38 days post-planting, measuring factors such as leaf water potential, osmotic potential, osmotic regulation, water consumption, water use efficiency, leaf freshness, leaf sodium content, and yield.

Effective management of electrical conductivity (EC) and pH levels is crucial for optimizing nutrient uptake and enhancing both productivity and quality in hydroponically grown leafy greens. Typically, the pH of nutrient solutions is maintained between 5.5 and 6.5; however, lowering the pH to a more acidic level (below 5) can significantly reduce many troublesome waterborne diseases. Plant responses to altered pH levels vary by species and can significantly influence the concentration of hydronium ions as well as perturb the availability of cations. A decrease in pH leads to an increase in hydronium ions, resulting in reduced availability of cations [66].

According to Purba et al. [85] studied that, in their experiment, various lettuce varieties were cultivated in different growth media, revealing a strong correlation among root wet weight, total wet weight, total oven dry weight, and root volume. The choice of growth medium had a significant impact on leaf area, crown oven dry weight, and root oven dry weight of the lettuce cultivars 30 days post-planting. However, neither the growing medium used nor the type of lettuce selected notably affected any plant parameters, such as length, number of leaves at 9, 16, and 23 days, shoot wet weight, leaf area, economic fresh weight, or leaf-root ratio.

Key parameters that typically require regulation include the pH of the growing medium, moisture levels, and ambient light and temperature conditions. While all parameters can be managed manually, sunlight is often substituted with artificial grow lights for indoor farming in urban environments. The types of grow lights available include light-emitting diode (LED) grow lights, fluorescent grow lights, and high-pressure sodium (HPS) or high-intensity discharge (HID) grow lights [86].

Various factors play a crucial role in hydroponic cultivation, including light intensity, quality, and duration; the composition and balance of nutrients; water quality; temperature and humidity; air circulation and ventilation; the growing medium; as well as pH and electrical conductivity levels. Additionally, plant variety and genetics, system design and maintenance, along with effective monitoring and control, are essential considerations [87].

Hydroponic cultivation offers numerous benefits, including the production of high-quality crops, improved health benefits, and significant cost savings, while also lowering labor costs. Crops cultivated through this method typically exhibit superior nutritional value, with vitamin levels being approximately 50% higher than those grown using traditional methods. Additionally, the runoff from conventional gardening can cause environmental harm due to elevated levels of dissolved calcium, phosphorus, and potassium. One enhancement of the traditional hydroponic approach involves managing water pH, a crucial factor for plant health. By maintaining optimal pH levels, plants can effectively absorb essential nutrients like calcium, magnesium, and boron [88].

2.4 Growing substrate-based hydroponic systems and their constituents

The development of composite substrates, which combine diverse materials to enhance physical and chemical properties, has opened up new possibilities for tailored solutions that cater to specific crop requirements. However, this increased versatility often comes at a dual cost: not only are advanced substrates typically more expensive than traditional options but they may also present complex management and maintenance challenges [16, 56]. As novel substrate materials emerge, they require growers to adapt their cultivation techniques and acquire new expertise to achieve optimal outcomes. While these substrates are touted as sustainable alternatives, it's crucial to consider the environmental implications of their production processes, including the potential for energy-intensive manufacturing and transportation, which may offset any eco-friendly benefits [80].

The most common substrate-based hydroponic system, for example, biochar is a renewable resource that enhances soil structure and retains nutrients effectively. It is created through pyrolysis at high temperatures in an environment with little to no oxygen. Biochar serves as both carbon storage and a soil fertilizer, and it can raise the pH level of soil while increasing its cation exchange capacity. In most research, "biochar" typically refers to "pyrochar," which is a by-product of processes such as fast pyrolysis or gasification. Conversely, hydrochar is produced via hydrothermal pyrolysis in high-pressure water. Due to its larger pore structure, biochar can better store water and nutrients compared to hydrochar. However, hydrochar possesses more oxygen functional groups, which increases its chemical reactivity within the soil [89], coir is a sustainable material with exceptional water retention and aeration properties. A study assessed the impact of four different commercial substrates, including a peat-based one and three types of coir (coir pith, coir chips, and a combination of coir pith and fibers), on the yield, phytochemical content, and antioxidant capacity of *Spinacia oleracea* L. cv. 'Manatee'. Spinach seedlings were cultivated in *styrofoam* planting boxes filled with these substrates. The results indicated that plants grown in these substrates achieved similar fresh yields but had a higher total flavonoid concentration compared to those grown in peat [90], Expanded clay pebbles: commonly referred to as "grow rocks," are produced by subjecting clay to extremely high temperatures. This synthetic product is available in various sizes, ranging from 1 to 18 mm, with the specific size chosen based on the desired level of aeration. It has a very low capacity for retaining water, which may necessitate more frequent watering. However, expanded clay boasts a high cation exchange capacity (CEC), enabling it to hold nutrients and salts effectively. Additionally, this growing medium can be cleaned, sterilized, and reused, contributing to its sustainability [91]. Rockwool: rock mineral wool (RMW) has significantly contributed to the advancement of hydroponics due to its fibrous nature and remarkable water retention capabilities. However, this has also led to a substantial increase in RMW waste, prompting research into various reuse strategies. This composting process and degradability of spent rock mineral wool from hydroponics (gRMW) when combined with varying ratios of biowaste compost. The research encompasses the examination of the physical and chemical properties of both the initial and final materials, as well as the potential ecological risks associated with the end product [92]: rice hulls: renewable, biodegradable, good aeration [90]; perlite: this material is characterized by its excellent ability to retain moisture and its substantial pore space, which enhances aeration and facilitates drainage. Additionally, perlite has a neutral pH and lacks cation exchange capacity (CEC). It is lightweight and sterile before application. While perlite maintains its structure without breaking down, it can be reduced to smaller fragments during handling and may generate dust, potentially leading to respiratory issues. Therefore, it is advisable to wear masks when dealing with large amounts. Perlite is frequently combined with other growing media for plant cultivation [91], Vermiculite: The soilless cultivation of plants frequently employs inorganic vermiculite as a growing medium. The physicochemical characteristics of expanded vermiculite-phlogopite products sourced from Kovdor in the Murmansk region of Russia have been used as hydroponic substrates for 15 years. Four different samples were analyzed: natural vermiculite, initially expanded vermiculite (treated by roasting at 700°C), spent expanded vermiculite, and spent vermiculite that was roasted again at the same temperature [93] which was depicted in (**Figure 1**) below.

Therefore, soilless farming, or hydroponics, is an innovative method that increases vegetable crop yields by providing organic nutrients directly to the plant roots through irrigation water, rather than relying on soil as a growth medium. Consequently, agriculture must embrace technologies that conserve water while enhancing the nutritional quality or biochemical properties of vegetable products. This method utilizes water more effectively than traditional farming practices [96].

2.5 Unique characteristics of hydroponic farming

The key benefits of hydroponics can be distilled as follows: minimizing the use of chemicals (fertilizers, pesticides, and growth promoters); eliminating the need for soil; and maximizing land utilization by achieving higher yields per unit area, reducing consumption, and improving water management. These advantages collectively result in a reduced ecological footprint, making hydroponics an appealing choice for controlled environment agriculture [82] which is depicted in **Table 3**.

The controlled environment in hydroponic systems allows for precise management of temperature, light, and nutrient levels, resulting in accelerated plant growth and higher yields compared to traditional farming methods. Moreover, hydroponics enables year-round cultivation, allowing crops to be grown in regions with unfavorable climates or limited arable land. The eco-friendly nature of hydroponics extends beyond water conservation and efficient land use, offering a promising solution for a more sustainable food system [21]. The growth of large-scale hydroponics facilities is expected to skyrocket in the coming years, driven by the widespread adoption of advanced technologies that control climate, lighting, and irrigation systems. These

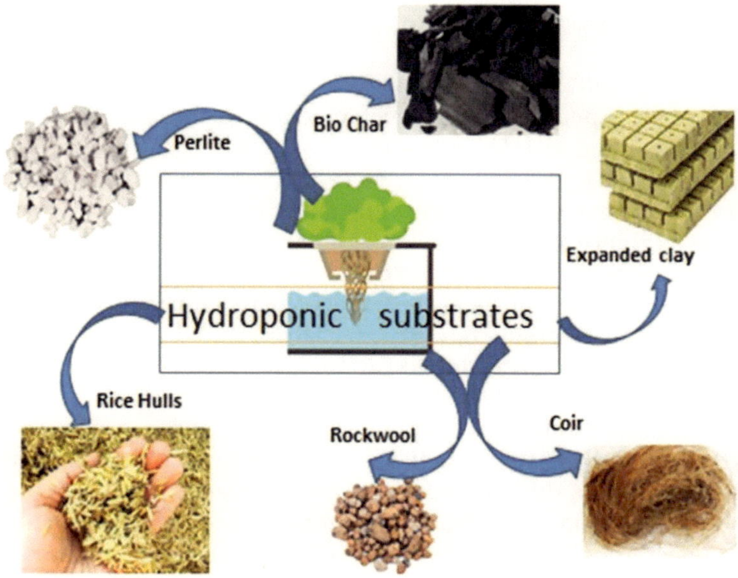

Figure 1.
Some substrate-based of hydroponic systems adapted from [88, 94, 95].

Conventional farming	Hydroponic cultivation/soilless farming	Reference
Soil-based farming	Soilless farming	[97]
Open field cultivation	Greenhouse cultivation	
Large cultivation area is needed	Optimal use of water	
Water wastage	Zero water wastage	
Required more space for cultivation	Required less space for cultivation	[63]
There is a higher wastage of fertilizer in soil-based farming	The nutrient solution contains only 25% of the essential elements required	
Over use of pesticides and chemicals, soil degradation, urbanization, natural disasters, and climate change	Its innovative farming techniques have emerged and growing crops without the use of soil	[98]
Not save land	For landless/save lands	[99]
Not optimized methods	Hydroponic technology continues to evolve, including advances in artificial intelligence and machine learning to optimize growing methods	
Not using Artificial intelligence (AI) can improve the accuracy of nutrient balancing, disease prediction, and environmental control, leading to increased efficiency and productivity	While using AI can improve the accuracy of nutrient balancing, disease prediction, and environmental control, leading to increased efficiency and productivity	[94]
Water not using properly	Water saving	[100]
Low-quality yield regardless of grown crops	High-quality yield regardless of grown crop; hydroponic systems could be a way to increase the food production sustainability in the future	
Open system	Closed system	[101]

Table 3.
The comparison between traditional and hydroponic cultivation in modern techniques.

technologies are made possible by the development of sophisticated sensors, online platforms, software, and mobile applications [16].

Hydroponics offers a promising solution to alleviate this strain by providing a sustainable alternative to traditional farming methods. One of the key advantages of hydroponics is its ability to reduce soil erosion, minimize water consumption, and optimize nutrient recycling, resulting in a more environmentally friendly approach. Furthermore, hydroponics has the potential to increase crop yields, making it an attractive option for food production [95].

Hydroponics has the potential to revolutionize the way it grows food, offering a sustainable and environmentally-friendly approach to agriculture. By eliminating the need for traditional soil-based farming, hydroponics reduces soil erosion and minimizes the use of chemicals, pesticides, and other environmentally harmful substances. Additionally, this method allows for the integration of eco-friendly pest control methods, creating a more holistic approach to sustainable agriculture [80]. Hydroponic systems have emerged as a valuable solution for urban farming, offering precise control over plant growth while requiring less space and resources than traditional methods. Urban areas often face challenges in finding suitable land for farming, making hydroponics a practical alternative [102] which is depicted in **Figure 2**.

Figure 2.
Unique characterics of hydroponic system in modern farming adapted from [53, 64, 103].

The hydroponic system offers a superior alternative to conventional soil-based cultivation, allowing crops to thrive in environments with poor soil quality. Hydroponic cultivation has been shown to produce high-quality crops while utilizing water resources up to 90% more efficiently. By providing a controlled and precise nutrient solution, this method enables optimal crop growth, resulting in increased yields. One of the most significant advantages of hydroponics is its water conservation capabilities, as the water used is not wasted through runoff or evaporation like in outdated farming methods. This sustainable approach to agriculture also reduces the environmental impact of farming, making it an attractive option for eco-conscious farmers [32]. Hydroponics can be ecologically beneficial, provided it is conducted in a sustainable way that considers the management of fertilizers, water, and materials. The significant growth of hydroponics over the past 20 years is largely attributed to the adoption of rock mineral wool (RMW) as a synthetic growing medium. Using artificial substrates in controlled environments minimizes the risk of crops being affected by pests, diseases, and weeds, resulting in increased yields [92].

2.6 Potential application of hydroponics farming system in modern agriculture

Hydroponic farming systems, particularly those utilizing vertically stacked growing methods, offer a promising alternative to traditional soil-based cultivation. By optimizing space utilization, these soilless systems can increase crop yields per unit area, making them an attractive solution for farmers looking to maximize production. This innovative approach can help farmers achieve greater yields while minimizing water consumption, making it an attractive option for sustainable agriculture practices [101].

Hydroponics, a form of soilless cultivation, has gained widespread popularity worldwide due to its remarkable water efficiency compared to traditional farming methods. The success of hydroponic agriculture can be attributed to its triple-fold benefits: enhanced productivity, improved safety, and optimized water resource management. At its core, hydroponics offers a unique advantage in terms of plant growth, allowing for faster and more controlled nutrient uptake. Additionally, the technology ensures a safer and healthier environment for crops by eliminating soil-borne pathogens and pests [104].

Rosa-Rodriguez et al. [105] studied that closed hydroponic systems are more efficient in terms of water and fertilizer consumption compared to open systems. A closed system was shown to produce 13.5 kg more tomatoes per cubic meter of water than an open system. This highlights the potential for optimizing water resources through greenhouse production, where environmental conditions can be precisely controlled, leading to more efficient crop growth. Sustainable agriculture, as a comprehensive approach, not only preserves the environment's integrity but also fosters long-term productivity and profitability by harmonizing three interconnected pillars: environmental stewardship, economic viability, and social responsibility. By adopting hydroponics, farmers can ensure a sustainable future by minimizing their ecological footprint, optimizing resource utilization, and promoting equitable access to healthy food for all [106].

It is valuable and essential to trace the beginnings of soilless growth and the evolution of elements like efficient growing media to place these technologies in the perspective of food production history. Simplified soilless systems are the emphasis of this section, as they have been actively promoted by organizations like the FAO to increase food security in underdeveloped nations [103].

In regions plagued by severe soil degradation and water scarcity, hydroponic farming offers a promising solution. This method of cultivation not only conserves water but also eliminates the need for pesticides, making it an environmentally friendly option. In areas ravaged by prolonged drought, where many struggle to afford water or dig wells, hydroponics provides a vital alternative to traditional farming practices, significantly reducing water consumption [63, 107].

In recent years, hydroponics has gained traction as a viable solution for cultivating a diverse range of crops. This method offers a sustainable approach to food production, eliminating the need for hazardous chemicals and ensuring compliance with strict environmental regulations [7]. The use of controlled environment agriculture (CEA) technologies, including vertical farming and hydroponics, is revolutionizing traditional farming practices by offering numerous benefits such as enhanced crop yields, reduced land and water usage, and year-round crop production. One such innovative approach is vertical farming, which involves stacking crops vertically and utilizing artificial lighting and precise nutrient delivery systems to optimize growth [108].

As the world's agricultural systems continue to evolve, hydroponic farming is gaining traction as a viable and sustainable option for food production. By leveraging greenhouses, hydroponics can provide a consistent supply of high-quality fruits and vegetables year-round, while minimizing resource consumption and reducing greenhouse gas emissions. This innovative approach not only benefits the environment but also empowers farmers to increase their competitiveness and differentiate themselves in the market [109].

The soilless cultivation method of hydroponics offers a promising solution to increase crop yields and sustainability, particularly in the face of decreasing arable land and water scarcity. By adopting decentralized precision methods, farmers can enhance stability and adapt to changing environmental conditions. While the technology has primarily been applied in profitable vegetable and herb markets in developed economies, expanding training programs and accessible technologies globally will be essential to widespread adoption and realization of its full potential [95].

2.7 Future trends for hydroponic farming system best practice in modern techniques

Hydroponic methods, especially the Nutrient Film Technique, are well-regarded for their ability to operate without soil and for their efficient circulation of nutrients and water. However, controlling optimal temperatures within a greenhouse can be difficult. One approach to minimize energy expenses involves managing the temperature of the nutrient solution rather than attempting to heat or cool the entire greenhouse space. By effectively regulating the temperature in the root zone through nutrient solution management, growers can create more suitable conditions for temperate leafy greens [31]. In the United States, certified organic production can utilize organic fertilizers within hydroponic systems. This approach serves as a basis for innovative technologies aimed at creating sustainable, circular food production systems that connect energy, water, and nutrient management [43].

Hydroponic agriculture is an innovative method of growing plants without soil, which has the potential to improve crop productivity and promote sustainability. As arable land and water resources continue to decline, hydroponics can serve as a complementary approach to traditional farming, helping to address global food security challenges. Innovations in automation, renewable energy use, biocontrol methods, and customized crop varieties have the potential to mitigate existing challenges. Advances in hydroponics and vertical farming can facilitate sustainable intensification to satisfy future nutritional requirements [82]. In hydroponics, plants thrive in a nutrient-rich solution rather than soil, allowing for optimal growth and development. This innovative method offers several benefits, including precise control over the levels of essential elements, efficient water usage, and the ability to cultivate crops in areas with subpar soil conditions [110]. By providing plants with a tailored blend of nutrients, hydroponics enables farmers to optimize growth rates, yields, and overall plant health [88].

The rising global population has led to an increase in food demand, highlighting the need for innovative and sustainable agricultural practices. Concerns about the depletion of fertile soil and the limited availability of agricultural land are becoming increasingly prominent. One potential breakthrough in addressing these issues could be the integration of soilless farming and microgreen cultivation with nanotechnology, which could offer a more sustainable and efficient alternative to traditional farming methods [111]. To mitigate the pressing need for sustainable agriculture, it is crucial to develop climate-resilient agricultural systems that prioritize resilience, resource conservation, and disease management. Hydroponics, a soilless cultivation method, has garnered significant attention due to its potential to reduce reliance on agricultural land and pesticides, as well as its capacity to thrive in areas with poor soil quality. This innovative approach can help alleviate the detrimental effects of extreme weather events and provide a more reliable food supply [112].

3. Conclusion

This chapter explores the use of hydroponic farming, a modern and sustainable cultivation technique that has gained popularity in recent years. The evidence presented in this paper highlights the potential applications of hydroponics, which involve growing plants in a nutrient-rich solution rather than soil.

Since this hydroponic farming method reduces pollutants, impacts ecosystems less, and mitigates climate change, it will contribute to global sustainability. The efficiency of hydroponic systems can be maximized by integration with building architecture, particularly in urban settings. It is best to include them in the early stages of new building design. Every technological, engineering, architectural, and infrastructure component should be taken into account throughout this integration. In areas with subpar soil quality or contamination, hydroponic cultivation offers a viable alternative. By providing a precise balance of nutrients, hydroponics promotes optimal growth and development, resulting in healthier and higher-quality crops. Recently, indoor hydroponic farming has emerged as a promising approach to ensure stable and increased food production, regardless of weather or location. Despite the progress made in developing hydroponic robot systems, there is still a need for more compact and cost-effective solutions for smaller indoor spaces. However, as research and technological advancements continue to improve the efficiency and affordability of hydroponics, it is expected that these systems will become more accessible and feasible for use in a variety of settings. This could lead to a significant shift in the agricultural industry, allowing for more sustainable and environmentally friendly practices to become the efficient use of resources, as well as continuous increases in production.

Acknowledgements

The authors are grateful to acknowledge all who had been instrumental in the creation of this chapter.

Authors' contributions

G.K. designed and validated the study and wrote the manuscript; A.A. reviewed and edited the manuscript; S.H. revised the manuscript; D.A. prepared the figures of this chapter. Also, all authors have approved the latest version of the manuscript and agree to be held accountable for the content therein.

Funding

No funding was received from any funding agency or organization toward making this manuscript.

Conflict of interest

The author declares that there are no conflicts of interest regarding the publication of this article.

Data availability

All data presented or analyzed during this study are included in this article.

Abbreviations

AI	artificial intelligence
CEA	controlled environment agriculture
DWC	deep water culture
NASA	National Aeronautics and Space Administration
NDVI	normalized difference vegetation index
NFT	nutrient film technique
HID	high-intensity discharge
IoT	Internet of Things
LED	light-emitting diodes
SDGs	Sustainable Development Goals.

Author details

Gamachis Korsa[1,2]*, Abate Ayele[1,2], Setegn Haile[1,2] and Digafe Alemu[1,2]

1 Department of Biotechnology, Center of Excellence for Biotechnology and Bioprocess, Addis Ababa Science and Technology University, Addis Ababa, Ethiopia

2 Center of Excellence for Biotechnology and Bioprocess, Addis Ababa Science and Technology University, Addis Ababa, Ethiopia

*Address all correspondence to: gamachis.korsa@aastu.edu.et

References

[1] Ghosh A, Kumar A, Biswas G. Exponential population growth and global food security: Challenges and alternatives. In: Bioremediation of Emerging Contaminants from Soils. Amsterdam: Elsevier; 2024. pp. 1-20. DOI: 10.1016/B978-0-443-13993-2.00001-3

[2] Wimmerova L, Keken Z, Solcova O, Bartos L, Spacilova MA. Comparative LCA of aeroponic, hydroponic, and soil cultivations of bioactive substance-producing plants. Basel, Switzerland: Sustainability. 2022;**14**:2421. DOI: 10.3390/su14042421

[3] Gorjian S, Calise F, Kant K, Ahamed MS, Copertaro B, Najafi G, et al. A review of opportunities for implementation of solar energy technologies in agricultural greenhouses. Journal Cleaner Production. 2021;**28**:124807. DOI: 10.1016/j.jclepro.2020.124807

[4] Pradhan B, Deo B. Soilless farming-the next generation green revolution. Current Science. 2019;**116**(5):728-732. Available from: https://www.jstor.org/stable/27137929

[5] Chaudhary A, Anand S. Soilless cultivation: A distinct vision for sustainable agriculture. In: Artificial Intelligence and Smart Agriculture: Technology and Applications. Singapore: Springer Nature; 2024. pp. 337-368. DOI: 10.1007/978-981-97-0341-8_17

[6] Sousa R, Bragança L, da Silva MV, Oliveira RS. Challenges and solutions for sustainable food systems: The potential of home hydroponics. Sustainability. 2024;**16**:817. DOI: 10.3390/su16020817

[7] Debangshi. Hydroponics-an overview. Indian Chronicle of Bioresource Management. 2021;**5**(3):110-114. Available from: https://pphouse.org/cbm.php

[8] Jain A, Kumari N, Jha VK. A review on the hydroponic system: Hope and hype. In: Chapter-11. Recent Advances in Chemical Sciences and Biotechnology. New Delhi Publication; 2019. pp. 143-149. Available from: https://www.researchgate.net/profile/Vikash-Jha-6/

[9] Payen FT, Evans DL, Falagán N, Hardman CA, Kourmpetli S, Liu L, et al. How much food can we grow in urban areas? Food production and crop yields of urban agriculture: A meta-analysis. UK: Earth's Future. 2022;**10**(8):1-22, e2022EF002748. DOI: 10.1029/2022EF002748

[10] Megha B, Preeti YH, Ravindra MM, PD K. Roots in water: The rise of hydroponics in modern farming. Agriculture Magazine. Junagadh, Gujarat: JAU; 2024;**3**(9):1-7. Available from: https://www.researchgate.net/profile/Ravindra-Muchhadiya/

[11] Jan S, Rashid Z, Ahngar TA, Iqbal S, Naikoo MA, Majeed S, et al. Hydroponics–a review. International Journal of Current Microbiology and Applied Science. 2020;**9**(8):1779-1787. DOI: 10.20546/ijcmas.2020.908.206

[12] Olubanjo OO, Adaramola OD, Alade AE. Development of greenhouse with root dipping technique hydroponics structure to test the performance of jute mallow. European Journal of Agriculture and Food Sciences. 2021;**3**(4):10-18. Available from: www.ejfood.org

[13] Resh HM. Hydroponic Food Production: A Definitive Guidebook for the Advanced Home Gardener and the Commercial Hydroponic Grower. Boca

Raton, Florida: CRC Press; 2022. pp. 642-649. DOI: 10.1201/9781003133254

[14] Putra PA, Yuliando H. Soilless culture system to support water use efficiency and product quality: A review. Agriculture and Agricultural Science Procedia. 2015;**3**:283-288. DOI: 10.1016/j.aaspro.2015.01.054

[15] Fuentes-Peñailillo F, Gutter K, Vega R, Silva GC. New generation sustainable Technologies for Soilless Vegetable Production. Horticulture (Basel, Switzerland: MDPI). 2024;**10**(1):49. DOI: 10.3390/plants11091153

[16] Velazquez-Gonzalez RS, Garcia-Garcia AL, Ventura-Zapata E, Barceinas-Sanchez JDO, Sosa-Savedra JC. A review of hydroponics and the technologies associated for medium- and small-scale operations. Agriculture (Basel, Switzerland). 2022;**12**:646-655. DOI: 10.3390/agriculture12050646

[17] Mishra SJ, Rout D, Sahoo D. Analysing the economic viability of hydroponic farming: A comparative cost-benefit analysis. International Journal of Progressive Research in Engineering Management and Science (IJPREMS). 2024;**4**(6):1806-1811. DOI: 10.58257/IJPREMS35128

[18] Gilmour DN, Bazzani C, Nayga RM Jr, Snell HA. Do consumers value hydroponics? Implications for organic certification. Agriculture Economics (Hoboken, NJ: Wiley-Blackwell). 2019;**50**:707-721. DOI: 10.1111/agec.12519

[19] Van Os EA, Gieling TH, Lieth JH. Technical equipment in soilless production systems. In: Soilless culture. Amsterdam, Netherlands: Elsevier; 2019. pp. 587-635. DOI: 10.1016/B978-0-444-63696-6.00013-X

[20] Mielcarek A, Kłobukowska K, Rodziewicz J, Janczukowicz W, Bryszewski KŁ. Water nutrient management in soilless plant cultivation versus sustainability. Sustainability (Basel, Switzerland). 2023;**16**(1):152. DOI: 10.3390/su16010152

[21] Kumar P, Singh J. Hydroponic: An eco-friendly future. In: Kumar N, editor. Hydroponics and Environmental Bioremediation. Springer Water. Cham: Springer; 2024. pp. 223-267. DOI: 10.1007/978-3-031-53258-0_9

[22] Paturu P, Varadarajan S. Assessing environmental sustainability by combining product service systems and life cycle perspective: A case study of hydroponic urban farming models in India. Science of The Total Environment. 2024;**927**:172232. DOI: 10.1016/j.scitotenv.2024.172232

[23] Hussain A, Iqbal K, Aziem S, Mahato P, Negi AK. A review on the science of growing crops without soil (soilless culture) a novel alternative for growing crops. International Journal of Agriculture and Crop Sciences (United State). 2014;**7**(11):833. Available from: https://www.researchgate.net/profile/Kaiser-Iqbal/

[24] Sharma N, Acharya S, Kumar K, Singh N, Chaurasia OP. Hydroponics as an advanced technique for vegetable production: An overview. Indian Society of Soil and Water Conservation. 2018;**17**(4):364-371. DOI: 10.5958/2455-7145.2018.00056.5

[25] Kluko DA. Home hydroponics. Popular Mechanics. Hearst Communications. 2017;**2**:85-87. Available from: http://pop.h-cdn.co/assets/downloads/1487018470_-_pmx020117project-lo.pdf

[26] Rani RS, Kumar HV, Mani A, Reddy BS, Rao CS. Soilless cultivation

technique, hydroponics—A review. Current Journal of Applied Science and Technology. 2022;**41**(13):22-30. DOI: 10.9734/cjast/2022/v41i1331711

[27] Giacomelli GA, van Weel PA, Blok C. Ebb and flood nutrient delivery system for sustainable automated crop production. Acta Horticulturae. 2020;**1296**:1129-1136. DOI: 10.17660/ActaHortic.2020.1296.143

[28] Lee S, Lee J. Beneficial bacteria and fungi in hydroponic systems: Types and characteristics of hydroponic food production methods. Scientia Horticulturae. 2015;**195**:206-215. DOI: 10.1016/j.scienta.2015.09.011

[29] Hamza A, Abdelraouf RE, Helmy YI, El-Sawy SM. Using deep water culture as one of the important hydroponic systems for saving water, and mineral fertilizers and improving the productivity of lettuce crops. International Journal of Health Sciences. 2022;**6**(59):2311-2331. DOI: 10.53730/ijhs.v6nS9.12932

[30] Santosh DT, Dinkar JG. Advances in hydroponic systems: Types and management. In: Sagar M, Dinkar JG, Santosh DT, editors. Advances in Agricultural Technology. Canada: Griffon; 2023. pp. 16-28. Available from: https://www.researchgate.net/profile/Dt-Santosh/

[31] Ilahi WF, Rosdi NS, Tajri AA. Evaluating temperature variation of nutrient solution in nutrient film technique (NFT) cooling system for temperate vegetable. Journal of Tropical Resources and Sustainable Science (JTRSS). 2024;**12**(1):47-50. DOI: 10.47253/jtrss.v12i1.1364

[32] Monisha K, Kalai Selvi H, Sivanandhini P, Sona Nachammai A, Anuradha CT, Rama Devi S, et al. Hydroponics agriculture is a modern agriculture technique. Journal of Achievements in Materials and Manufacturing Engineering. 2023;**116**(1):25-35. DOI: 10.5604/01.3001.0016.3395

[33] Hayati R, Armadi Y, Yusnaweti Y, Feni R, Asfaruddin A, Ariyani F, et al. Testing alternative hydroponic nutrition of Wick system on vegetable growth planted in limited land. Journal La Lifesci. 2024;**5**(2):94-108. DOI: 10.37899/journallalifesci.v5i2.1177

[34] Georgios S, Savvas D, Agyropolou K, Tarantillis PA. Yield and nutritional quality of aeroponically cultivated basil as affected by the available root-zone volume. Emirates Journal of Food and Agriculture. 2015;**27**(12):911-918. DOI: 10.9755/ejfa.2015-05-233

[35] Dubey N, Nain V. Hydroponic-the future of farming. International Journal of Environment, Agriculture and Biotechnology. 2020;**5**(4):1-8. DOI: 10.22161/ijeab.54.2

[36] Opačić N, Radman S, Fabek Uher S, Benko B, Voća S, Šic ŽJ. Nettle cultivation practices from open field to modern hydroponics: A case study of specialized metabolites. Plants (Basel, Switzerland). 2022;**11**(4):1-19. DOI: 10.3390/plants11040483

[37] Kaushik K, Tare K. Aeroponics farming system: A step towards modern agriculture. Food and Scientific Reports. Jorhat. 2024;**5**(5):48-52. Available from: foodandscientificreports.com

[38] Chittibomma K, Yadav NK, Reddy MG. Aeroponics: A polytropic research tool in the new era of agriculture. International Journal of Environment and Climate Change. 2023;**13**:214-218. DOI: 10.9734/IJECC/2023/v13i81946

[39] Akinola TE, Lawal PO. Development of aeroponics stands using locally sourced materials. Iconic Research and Engineering Journals. 2023;**7**(1):19-23. Available from: https://www.irejournals.com/formatedpaper/1704831.pdf

[40] Joly A, Junge R, Bardocz T. Aquaponics business in Europe: Some legal obstacles and solutions. Ecocycles. 2015;**1**(2):3-5. DOI: 10.19040/ecocycles.v1i2.30

[41] Gunning D, Maguire J, Burnell G. The development of sustainable saltwater-based food production systems: A review of established and novel concepts. Water. 2016;**8**(12):598. DOI: 10.3390/w8120598

[42] Miličić V, Thorarinsdottir R, Dos SM, Turnšek HM. Commercial aquaponics approaching the European market: To consumers' perceptions of aquaponics products in Europe. Water. 2017;**9**(2):80. DOI: 10.3390/w9020080

[43] Park Y, Williams KA. Organic hydroponics: A review. Scientia Horticulture. 2024;**324**:112604. DOI: 10.1016/j.scienta.2023.112604

[44] Ghimire A, Rojan Karki MD. Hydroponics: An innovative approach to urban agriculture. Nepalese Journal of Agricultural Sciences. 2023;**89**:92-98. Available from: http://www.nepjas.com/uploads/article/pdf/

[45] Van Gorcum B, Goddek S, Keesman KJ. Gaining market insights for aquaponically produced vegetables in Kenya. Aquaculture International. 2019;**27**:1231-1237. DOI: 10.1007/s10499-019-00379-1

[46] Vlăduț NV, Ungureanu N. Beyond agriculture 4.0: Design and development of modern agricultural machines and production systems. Agriculture. 2024;**14**(7):1-4. DOI: 10.3390/agriculture14070991

[47] Sadek N, Shehata D. Internet of things based smart automated indoor hydroponics and aeroponics greenhouse in Egypt. Ain Shams Engineering Journal. 2024;**15**(2):102341. DOI: 10.1016/j.asej.2023.102341

[48] Chowdhury H, Asiabanpour B. Influencing factors for the plant growth patterns in hydroponic and aquaponic environments: A subgroup analysis for sustainable agricultural production. Green Technologies and Sustainability. 2024;**2**(2):1-15. DOI: 10.1016/j.grets.2024.100084

[49] Sharma N, Acharya S, Kumar K, Singh N. Hydroponics as an advanced technique for vegetable production – An overview. Journal of Soil and Water Conservation. 2019;**17**(4):364-371. DOI: 10.5958/2455-7145.2018.00056.5

[50] Zhang L, Yang J, Li D, Liu H, Xie Y, Song T, et al. Evaluation of the ecological civilization index of China based on the double benchmark progressive method. Journal of Cleaner Production. 2019;**222**:511-519

[51] Mittal R, Bhukal S. Sustainable approach for agriculture and environmental remediation using hydroponics and their perspectives. In: Hydroponics and Environmental Bioremediation: Wastewater Treatment. Cham, Switzerland: Springer Nature; 2024. pp. 65-90. DOI: 10.1007/978-3-031-53258-0_3

[52] Pomoni DI, Koukou MK, Vrachopoulos MG, Vasiliadis L. A review of hydroponics and conventional agriculture based on energy and water consumption, environmental impact, and land use. Energies. 2023;**16**:1690. DOI: 10.3390/en16041690

[53] Rajendran S, Domalachenpa T, Arora H, Li P, Sharma A, Rajauria G. Hydroponics: Exploring innovative sustainable technologies and applications across crop production, with emphasis on potato mini-tuber cultivation. Heliyon. 2024;**10**(5):e26823. DOI: 10.1016/j.heliyon.2024.e26823

[54] Kumar P, Sampath B, Kumar S, Babu BH, Ahalya N. Hydroponics, aeroponics, and aquaponics technologies in modern agricultural cultivation. In: Trends, Paradigms, and Advances in Mechatronics Engineering. Hershey, PA, USA: IGI Global; 2023. pp. 223-241. DOI: 10.4018/978-1-6684-5887-7.ch012

[55] Fayezizadeh MR, Ansari NA, Albaji M, Khaleghi E. Effects of hydroponic systems on yield, water productivity and stomatal gas exchange of greenhouse tomato cultivars. Agricultural Water Management. 2021;**258**:107171. DOI: 10.1016/j.agwat.2021.107171

[56] Lakhiar G, Jianmin TN, Syed FA, Chandio NA, Buttar WA, Qureshi. Monitoring and control systems in agriculture using intelligent sensor techniques: A review of the aeroponic system. Journal of Sensors. 2018;**2018**(1):8672769. DOI: 10.1155/2018/8672769

[57] Kumari R, Kumar R. Aeroponics: A review on modern agriculture technology. Indian Farmer. 2019;**6**(4):286-292. Available from: https://www.researchgate.net/profile/Ramesh-Kumar-Bhardwaj-2/

[58] Sohi A, Attri T, Kaur G, Dogra BS. Vertical Farming: Revolutionizing Urban Agriculture. New Delhi: Integrated Publications TM; 2023. pp. 235-250. Available from: https://www.researchgate.net/profile/

[59] Mahkeswaran R, Ng AK, Toh C, Toh B. Multi-loop aquaponics systems: A review and proposed multi-loop agrogeological aquaponics system. Journal of Advanced Agricultural Technologies. 2024;**11**(1):1-7. DOI: 10.18178/joaat.11.1.1-7

[60] Zhao W, Liu C, Li D, Zheng Y. Energy-saving techniques in urban aquaponics farms by optimizing equipment operating scheme. Aquaculture (Amsterdam, Netherlands). 2024;**587**:740873. DOI: 10.1016/j.aquaculture.2024.740873

[61] Maitland DM, Savidov NA, Shultz RC, Colombo SM. Aquaponics systems: A holistic, ecosystem approach to food production. In: Future Food Systems. Amsterdam, Netherlands: Academic Press; 2024. pp. 147-160. DOI: 10.1016/B978-0-443-15690-8.00024-2

[62] Kotzen B, Emerenciano MG, Moheimani N, Burnell GM. Aquaponics: Alternative types and approaches. In: Aquaponics Food Production Systems: Combined Aquaculture and Hydroponic Production Technologies for the Future. Cham, Switzerland: Springer; 2019. pp. 301-330. DOI: 10.1007/978-3-030-15943-6

[63] Son JE, Kim HJ, Ahn TI. Hydroponic systems. In: Plant Factory. Academic Press; 2020. pp. 273-283. DOI: 10.1016/B978-0-12-816691-8.00020-0

[64] Jain S, Kaur M. Design and implementation of an IoT-based automated EC and pH control system in an NFT-based hydroponic farm. Engineering, Technology and Applied Science Research. 2024;**14**(1):13078-13081

[65] Al Mamun A, Naznen F, Jingzu G, Yang Q. Predicting the intention and

adoption of hydroponic farming among Chinese urbanites. Heliyon. 2023;**9**(3). DOI: 10.1016/j.heliyon.2023.e14420

[66] Gillespie DP, Kubota C, Miller SA. Effects of low pH of hydroponic nutrient solution on plant growth, nutrient uptake, and root rot disease incidence of basil (*Ocimum basilicum* L.). Horticulture Science. 2020;**55**(8):1251-1258. DOI: 10.21273/HORTSCI14986-20

[67] Hopkinson S, Harris M. Effect of pH on hydroponically grown bush beans (*Phaseolus vulgaris*). International Journal of Environment Agriculture and Biotechnology. 2019;**4**(1):142-145

[68] Savvas D, Gruda N. Application of soilless culture technologies in the modern greenhouse industry. A review. European Journal of Horticultural Science (Freiburg, Germany European). 2018;**83**(5):280-293. DOI: 10.17660/eJHS.2018/83.5.2

[69] Kulkarni BP, Mohite AS, Pai SV, Yedake PR. Automation of hydroponics system. International Journal of Science Technology and Engineering. 2017;**3**(9):600-602

[70] Taki M, Soheili-Fard F, Rohani A, Chen G, Yildizhan H. Life cycle assessment to compare the environmental impacts of different wheat production systems. Journal of Cleaner Production. 2018;**197**:195-207

[71] Hesampour R, Taki M, Fathi R, Hassani M, Halog A. Energy-economic-environmental cycle evaluation comparing two polyethene and polycarbonate plastic greenhouses in cucumber production (from production to packaging and distribution). Total Environmental Science. 2022;**828**:154232

[72] Hatfield JL, Prueger JH. Temperature extremes: Effect on plant growth and development. Journal of Weather and Climate Extremes. 2015;**10**:4-10. DOI: 10.1016/j.wace.2015.08.001

[73] Baddadi S, Bouadila S, Ghorbel W, Guizani A. Autonomous greenhouse microclimate through hydroponic design and refurbished thermal energy by phase change material. Journal of Cleaner Production. 2019;**211**:360-379. DOI: 10.1016/j.jclepro.2018.11.192

[74] Palande V, Zaheer A, George K. Fully automated hydroponic system for indoor plant growth. Procedia Computer Science. 2018;**129**:482-488. DOI: 10.1016/j.procs.2018.03.028

[75] Baxter LL, Grey TL, Tucker JJ, Hancock DW. Optimizing temperature requirements for clover seed germination. Agrosystems, Geosciences and Environment. 2019;**2**(1):1-7. DOI: 10.2134/age2018.11.0059

[76] Viršilė A, Olle M, Duchovskis P. LED lighting in horticulture. In: Light Emitting Diodes for Agriculture: Smart Lighting. Singapore: Springer; 2017. pp. 113-147. DOI: 10.1007/978-981-10-5807-3_7

[77] Yang J, Sun J, Wang X, Zhang B. Light intensity affects growth and nutrient value of hydroponic barley fodder. Agronomy. 2024;**14**(6):1-17. DOI: 10.3390/agronomy14061099

[78] Kim D, Son JE. Adding far-red to red, blue supplemental light-emitting diode inters lighting improved sweet pepper yield but attenuated carotenoid content. Frontiers in Plant Science (Lausanne, Switzerland). 2022;**13**:938199. DOI: 10.3389/fpls.2022.938199

[79] Zhou J, Chen J, Chen X, Zhu X, Qiu Y, Song H, et al. Sensitivity of six typical spatiotemporal fusion methods

to different influential factors: A comparative study for a normalized difference vegetation index time series reconstruction. Remote Sensing of Environment. 2021;**252**:112130. DOI: 10.1016/j.rse.2020.112130

[80] Rajaseger G, Chan KL, Tan KY, Ramasamy S, Khin MC, Amaladoss A, et al. Hydroponics: Current trends in sustainable crop production. Bio Information. 2023;**19**(9):925. DOI: 10.6026/97320630019925

[81] Kwon MJ, Hwang Y, Lee J, Ham B, Rahman A, Azam H, et al. Waste nutrient solutions from full-scale open hydroponic cultivation: Dynamics of effluent quality and removal of nitrogen and phosphorus using pilot-scale sequencing batch reactor. Journal of Environmental Management. 2021;**281**:111893

[82] Naresh R, Jadav SK, Singh M, Patel A, Singh B, Beese S, et al. Role of hydroponics in improving water-use efficiency and food security. International Journal of Environment and Climate Change. 2024;**14**(2):608-633. Available from: https://www.sdiarticle5.com/review-history/112351

[83] Zamora-Izquierdo MA, Santa J, Martínez JA, Martínez V, Skarmeta AF. Smart farming IoT platform based on edge and cloud computing. Biosystems Engineering. 2019;**177**:4-17. DOI: 10.1016/j.biosystemseng.2018.10.014

[84] de Carvalho Leal LY, de Souza ER, Júnior JA, Dos Santos MA. Comparison of soil and hydroponic cultivation systems for spinach irrigated with brackish water. Scientia Horticulturae. 2020;**274**:109616. DOI: 10.1016/j.scienta.2020.109616

[85] Purba JH, Parmila IP, Dadi W. Effect of soilless media (hydroponic) on growth and yield of two varieties of lettuce. Agricultural Science. 2021;**4**(2):154-165. Available from: http://agriculturalscience.unmerbaya.ac.id/index.php/agriscience/article/view/58

[86] Hemathilake DM, Gunathilake DM. High-productive agricultural technologies to fulfill future food demands: Hydroponics, aquaponics, and precision/smart agriculture. In: Future Foods. Academic Press; 2022. pp. 555-567. DOI: 10.1016/B978-0-323-91001-9.00018-9

[87] Kailashkumar B, Priyadharshini K, Logapriya M. Hydroponic cultivation: Factors affecting its success and efficacy. International Journal of Environmental and Climate Change. 2023;**13**(10):2403-2410. DOI: 10.9734/IJECC/2023/v13i102905

[88] Mujtahidin MH, Shah AF, Jais AS, Annuar KA, Sapiee MR. Design and development of control and monitoring hydroponic system. International Journal of Reconfigurable and Embedded Systems (IJRES). 2024;**13**:41-51. Available from: http://ijres.iaescore.com

[89] Eskandari S, Mohammadi A, Sandberg M, Eckstein RL, Hedberg K, Granström K. Hydrochar-amended substrates for production of containerized pine tree seedlings under different fertilization regimes. Agronomy. 2019;**9**(7):1-17. DOI: 10.3390/agronomy9070350

[90] Machado RM, Alves-Pereira I, Ferreira R, Gruda NS. Coir, an alternative to peat-effects on plant growth, phytochemical accumulation, and antioxidant power of spinach. Horticulture. 2021;**7**(6):1-6. DOI: 10.3390/horticulturae7060127

[91] Pinkerton M, Whitman B, Wooten H, Gomez C. Common media

used in hydroponics: EP623/ENH1359, 8/2022. EDIS. The Institute of Food and Agricultural Sciences (IFAS) and University of Florida; 2022. pp. 1-5. DOI: 10.32473/edis-EP623-2022

[92] Istenič D, Prosenc F, Zupanc N, Turel M, Holobar A, Milačič R, et al. Composting of recovered rock wool from hydroponics for the production of soil amendment. Environmental Science and Pollution Research. 2024;**31**(20):29280-29293. DOI: 10.1007/s11356-024-33041-2

[93] KremenetskayaI, IvanovaL, ChislovM, Zvereva I, Vasilieva T, Marchevskaya V, et al. Physicochemical transformation of expanded vermiculite after long-term use in hydroponics. Applied Clay Science. 2020 Nov;**15**(198):105839. DOI: 10.1016/j.clay.2020.105839

[94] Benko B, Uher SF, Radman S, Opačić N. Hydroponic production systems in greenhouses. In: Climate Smart Greenhouses-Innovations and Impacts. London, UK: IntechOpen; 2023. pp. 1-27. DOI: 10.5772/intechopen.113056

[95] Shin KK, Ping TP, Ling MG, Jiun CC, Bolhassan NA. Smart grow–low-cost automated hydroponic system for urban farming. HardwareX. 2024;**17**:e00498. DOI: 10.1016/j.ohx.2023.e00498

[96] Shilpa S, Sharma P, Bansuli B. Hydroponics in vegetable crops: A review. In: Hydroponics: The Future of Sustainable Farming. 2024. pp. 15-41. DOI: 10.3390/en16041690

[97] Kundal A, Sharma TP. Detection of plant diseases in hydroponics farming using deep learning techniques. In: 2024 International Conference on Smart Systems for Applications in Electrical Sciences (ICSSES). IEEE; 2024. pp. 1-6. DOI: 10.1109/ICSSES62373.2024.10561290

[98] Goddek S, Keesman KJ. Improving nutrient and water use efficiencies in multi-loop aquaponics systems. Aquaculture International. 2020;**28**(6):2481-2490. DOI: 10.1007/s10499-020-00600-6

[99] Verma RC, Chaithanya G, Sachan P, Thilagam P, Bahadur R, Kumar N, et al. Exploring hydroponics and the associated technologies for use in medium-and small-scale operations: A review. International Journal of Environment and Climate Change. 2023;**13**(10):4474-4483. DOI: 10.9734/ijecc/2023/v13i103125

[100] Okasha AM, Eldib EM, Elmetwalli AH, Farooque AA, Yaseen ZM, Elsayed S. Maximization of water productivity and yield of two iceberg lettuce cultivars in hydroponic farming system using magnetically treated saline water. Agriculture. 2022;**12**(1):101. DOI: 10.3390/agriculture12010101

[101] Indurthi S, Sarma I, Maheswarareddy C. Vegetable production through aquaponics: A prospective of future crop production. International Journal of Research in Agronomy. 2024;7(4):162-167. DOI: 10.33545/2618060X.2024.v7.i4c.553

[102] Farvardin M, Taki M, Gorjian S, Shabani E, Sosa-Savedra JC. Assessing the physical and environmental aspects of greenhouse cultivation: A comprehensive review of conventional and hydroponic methods. Sustainability. 2024;**16**(3):1273. DOI: 10.3390/su16031273

[103] Arumugam T, Sandeep G, Maheswari MU. Soilless farming of vegetable crops: An overview. Pharma Innovative Journal. 2021;**10**(1):773-785. Available from: www.thepharmajournal.com

[104] De la Rosa-Rodríguez R, Lara-Herrera A, Trejo-Tellez LI, Padilla-Bernal LE, Solís-Sánchez LO, Ortis-Rodríguez JM. Water and fertilizers productivity in closed and open hydroponic systems of tomato production. Horticultura Brasileira. 2020;**38**(1):47-52. DOI: 10.1590/S0102-053620200107

[105] Liao PA, Liu JY, Sun LC, Chang HH. Can the adoption of protected cultivation facilities affect farm sustainability? Sustainability. Basel, Switzerland. 2020;**12**(23):1-17

[106] Caputo S. History, techniques and technologies of soil-less cultivation. In: Small Scale Soil-less Urban Agriculture in Europe. Urban Agriculture. Cham: Springer; 2022. pp. 45-86. DOI: 10.1007/978-3-030-99962-9_4

[107] Maurya PK, Karde RY, Bayskar AS, Charitha N, Rai S. Innovative technologies such as vertical farming and hydroponics to grow crops in controlled environments. Advanced Farming Technology. 2023;**84**. Available from: https://www.sciencedirect.com/science/article/abs/pii/S2589014X23000993

[108] Nirbita P, Chan KY, Thien GS, Lee CL. Smart hydroponic farming system integrated with LED grow lights. Pertanika. Journal of Science and Technology. 2024;**32**(2):685-701. DOI: 10.47836/pjst.32.2.11

[109] Michalis E, Giatra CE, Skordos D, Ragkos A. Assessing the different economic feasibility scenarios of a hydroponic tomato greenhouse farm: A case study from Western Greece. Sustainability. Basel, Switzerland: MDPI. 2023;**15**(19):14233. DOI: 10.3390/su151914233

[110] Manos DP, Xydis G. Hydroponics: Are we moving towards that direction only because of the environment? A discussion on forecasting and a systems review. Environmental Science and Pollution Research. 2019;**26**(13):12662-12672. DOI: 10.1007/s11356-019-04933-5

[111] Maluin FN, Hussein MZ, Nik Ibrahim NN, Wayayok A, Hashim N. Some emerging opportunities of nanotechnology development for soilless and microgreen farming. Agronomy. 2021;**11**(6):1213. DOI: 10.3390/agronomy11061213

[112] Saikanth K, Singh BV, Sachan DS, Singh B. Advancing sustainable agriculture: A comprehensive review for optimizing food production and environmental conservation. International Journal of Plant and Soil Science. 2023;**35**(16):417-425. Available from: https://www.sdiarticle5.com/review-history/101939

Chapter 6

Perspective Chapter: Growing Berries in Substrate

Haifa Ben Hadj Daoud, Marco Rosario Butera and Jorge Pedro Rosa Duarte

Abstract

Berries such as strawberries, blueberries and raspberries have expanded into non-traditional growing areas worldwide. In recent years, the berry production system has changed to cultivation in substrate and pots and this alternative system has shown the potential to produce high yields of high-quality berries. The quality of the substrate and the parameters of the pot are known to influence plant growth and fruit production. Many authors have evaluated the composition of the growing media and the size of the container with the aim of optimizing the appropriate growing conditions that affect the plant growth and berry yield. The physical characteristics of the substrate play a special role in berry cultivation; then, a good substrate should be of high quality, in order to provide the plant root system with the adequate environment to develop. The management of water irrigation and nutrient contents during berry production in substrates has a great importance on the yield and quality of fruits. Thus, suitable irrigation systems and a fertilizer solution containing all the essential nutrients are essential for yield and high-quality berries. In this perspective, the objective of this chapter is to present the equipment and practices used for the cultivation of berries in substrate.

Keywords: strawberry, blueberry, raspberry, substrate, water management, nutrient management

1. Introduction

Berries are grown in all parts of the world with various geographic areas having their own unique varieties [1]. They have gained economic importance, and their cultivation is more widespread every year. The demand for high-quality berries is increasing globally [2], thereby encouraging growers to embrace new production techniques and varietal development(s) [1].

Strawberries (*Fragaria* × *ananassa* Duch) are among the most popular fruit in the world for their taste, aroma and high nutritional value offering numerous health benefits [3]. Strawberries are grown in various locations and climates worldwide, ranging from open fields to greenhouses. Strawberries are grown in various locations and climates worldwide, ranging from open fields to greenhouses. The main strawberry producers are China, USA, Mexico, Turkey, Spain, Egypt, Republic of Korea, Russia, Poland and Japan. China is the leading producer with the primary position in the world with an important annual production of 2.2 million tons [4]. The commercial strawberry industry has been improved by new production systems and management practices as well as breeding that play an important role in strawberry yield and quality.

Recently, interest in growing media has gained importance as a sustainable alternative to traditional soil cultivation methods [5]. In Europe, the ban of many chemicals adopted to sterilize the soils forced the growers to look for soilless solutions to escape the increasing disease pressure in the soils. For example, in 2017, in the United Kingdom, the majority of strawberries are grown on substrate (55% of production) and fertigation is provided by dripper irrigation system. The rest of the strawberry production (45%) is still in the soil [6]. The success of strawberry cultivation depends largely on the quality of the growing media used and nutrient management, which is a key factor to ensure high yields with sanitary quality [7]. Strawberry production faces several threats such as extreme weather conditions, pests and diseases. Therefore, many technological innovations have been adopted to modify production systems to improve strawberry yield and quality. Nevertheless, strawberry production has faced many challenges that have put pressure on research teams as well as private stakeholders across the world, to develop new mechanisms of adaptation to meet the increasing demand for high-quality strawberry production [8].

Blueberry (*Vaccinium sp. L.*) has become an important fruit crop worldwide, and the world blueberry production has increased significantly over recent years due to increasing demand for this healthy fruit [9]. For example, in the USA, blueberries rank only second to strawberries in popularity of berries [10]. This genus contains more than 400 species, of which 40% are native to Southeast Asia, 25% belong to North America and 10% belong to Central and South America. Several innovative production systems have been explored globally for this crop [11]. The substrates for growing this crop are increasing in importance [12] and their production is increasing worldwide, especially in regions with mild winter climates [13]. In substrates, blueberries have high productivity from the first year of cultivation. Planting blueberries in pots with selected substrate composition can be a good alternative to soil-based blueberry cultivation especially when fertilization and irrigation are well controlled [14], as well the soil characteristics do not allow to maintain the crop on a normal performance, where soil is with a high percentage of clay particles (>20%).

Raspberries (*Rubus idaeus* L.) are also one of the most popular fruits, in addition to their delicious taste, color and aroma, they offer a multitude of health benefits thanks to their richness in vitamins, minerals and antioxidants [15]. Raspberry production has steadily increased due to new production methods under a protected environment. Then, it has been cultivated in many regions of the world for several centuries. For example, red raspberry is commonly cultivated in at least 30 countries [16]. In Europe, the combination of different methods such as protected cultivation, substrate growing, using different plant types and innovation of new cultivars gives the possibility to produce excellent qualities from May to November [17]. Moreover, soilless cultivation of raspberry is becoming widely adopted, making the selection substrate and the cultivation management essential to reach optimal yield and quality [18].

For cultivation of berries, a great number of growing media with mixtures from different origins exist but depending on cultivar, cultivation system and also environmental conditions, yield and quality of berries can be affected, which requires a great management control. Therefore, further studies on cultivation of berries in substrates are needed. In this context, this chapter presents a comprehensive review of cultivation of berries in substrate using existing information that helps berries producers to understand such cultivation technique.

2. Growing strawberry in substrates

Strawberry (*Fragaria × ananassa Duch*) is one of the most important widely consumed small fruits in the world [3]. Global strawberry production has increased in acreage in the last decades due to the increasing demand. Strawberry plants grown in non-saline soils, slightly acid and with a good drainage system to prevent fungal diseases attacking the roots and crown of the plant [19]. Being a shallow-rooted plant strawberry needs effective nutrient management [3]. Traditional strawberry cultivation, which uses soil as a growing media, has several problems, including soil-transmitted diseases caused by *Verticillium* spp. and *Phytophthora* spp., nematodes, and soil-limiting factors [20], all of which result in severe production losses in commercial strawberry cultivation [21].

Many factors contribute to improve yield of strawberries such as production systems, cultivars and management practices by controlling weeds, extreme weather, pests and diseases. Therefore, there is an increased interest in controlled environment production in greenhouses that insure a great control of production management leading to increase yield and fruit quality [22]. In strawberry-producing area, most strawberries are grown in open-field conditions but in temperate regions such as northern and central Europe, Korea, Japan and some areas of China, strawberries are grown in greenhouses under soilless cultivation. In Europe, the Netherlands produces strawberries in glasshouses, with the controlled climate and irrigation, and CO_2 supply. In other countries, strawberries are grown in polyethylene-covered greenhouses, including micro- or macrotunnels, using various growing containers with different sustainable substrate compositions [3]. Traditionally strawberry soilless culture is done in peat substrate, but the use of other substrates is rising in different parts of the world to act as suitable substitutes for soil [23]. Various substrates are reported to be effective against soil pests as well as improving fruit quality including physical appearance of strawberry fruit such as weight and polar-equatorial diameter that are between the main selection criteria of consumers [24]. The improvement of these criteria could increase the value of commercial strawberry plantations for exportation [25]. The most growing media used for strawberries are perlite, cocopeat, peat moss and rockwool. Perlite treatment was found to have a positive effect on the improvement of fruit size, yield and fruit quality [21]. Perlite treatment was found to have a positive effect on the improvement of fruit size, yield and fruit quality [21]. Substrates with perlite and other mixtures have shown a great effect on root development by improving aeration insuring an effective nutrient uptake by shoot resulting in an increased yield [21]. For example, strawberry varieties, cv. Sweet Charlie and Camarosa when cultivated in perlite and its mixtures, showed an increased yield [26, 27]. Substrate has also an effect on strawberry flowering as reported by Anagnostou et al. [28] who revealed an early flower production in strawberry varieties of Fern and Selva grown in perlite under greenhouse conditions. Other studies showed the efficacy of combining perlite: peat-moss with mineral solutions in the improvement of strawberry vegetative growth [7, 29]. Perlite has been reported to be able to provide essential micronutrients to plants when mixed with manure and this is explained by the increase in berry size observed on strawberries in some studies [21, 30]. Similar results were reported by Linardakis and Manios [31] and Paraskevopoulou et al. [32], on the strawberry variety Selva grown in a perlite mixture (perlite medium + peat) showing maximum yield and berry weight compared to normal soil solution. The combination of perlite with farmyard manure (FYM) was reported by Thakur and Shylla [21] to be very effective and can be successfully used to produce good quality strawberry under polyhouse, increasing the flowering, berry set, berry size, weight and berry yield of strawberry. Coco fiber substrates (coco fiber and its mixtures) have also been reported as an excellent growing media for strawberries, which have been recommended to increase yield due to their high-water retention and good aeration [33, 34]. Other authors found that the best physical values of strawberry fruit were observed when using a combination of coconut fiber with other materials high in porosity [25, 35]. Those observations were confirmed with the findings of García-López and Cruz-Ortega [36] where the mixture of coconut fiber and volcanic rock showed the highest physical values in strawberry such as fruits per plant and fruit weight, compared using coconut fiber alone. Some studies showed a rapid compaction of coconut fiber as in the case of soil, while strawberry plants depend greatly on their roots' density [37]. However, the results of Rivera-del Rio et al. [38] on the hydroponic production of strawberries demonstrated that the substrates contain coconut fibers, small particles that show a large change in size, reducing the space for rooting, which decreases the growth of strawberries. The same study revealed that, unlike coconut fiber, when volcanic rock is used as a substrate, it increases water availability and maintains bulk density, porosity and aeration capacity, which positively impacts strawberry production. In Mexico, volcanic rock "Tezontle" is used as substrate for strawberry cultivation, which is very convenient for this crop not only for its low cost but also for improvement in physical fruit qualities. Tezontle is abundant in Mexico, and it is considered inert material with pH values near neutrality, low conductivity, good aeration and moisture-holding capacity [36].

Substrates not only affect strawberry growth and yield, but also fruit quality characteristics such as sugar content [21], and many authors have reported a positive influence of substrate mixtures on fruit sugar content [39, 40]. According to Sharma et al. [41], substrate composition can convert higher amounts of organic acids and photosynthesis into sugars during the fruit ripening phase, which affects berry quality. The increase in sugar content of berries could be explained by the acceleration of growth and fruiting and could also be due to some physiological processes such as carbohydrate synthesis in the plant system [42]. Contrary to these studies, García-López and Cruz-Ortega [36] reported in a recent work that the Brix degrees of strawberries were not affected using the substrate, whereas Ulrich and Olbricht [43] demonstrated that any increment in sugar content is correlated with lower production yields. The acidity of strawberries has also been assessed in some studies; Sharma [44] and Thakur and Shylla [21] showed that strawberries grown in soilless substrate ripened earlier than those grown in soil, which require a long period to reach the harvest stage. This is confirmed by Davis [45] and Hassan et al. [46] who found that strawberries grown in coir are less acidic than those grown in soil.

2.1 Container effect on strawberry cultivation in substrates

Several studies have been carried out on the cultivation of strawberries using different container types with different volumes, shapes (height) and colors (black, white). The size and shape of the container interacts with plant biomass and its allocation [47], and therefore has consequences for plant performance. In this context, the interactions between substrate and container size have been evaluated by many authors [48]. According to Sharma et al. [41], the use of growing media with different proportions in pots of different sizes significantly influences the canopy area, growth, yield and quality indicators [41]. This agrees with Manole et al. [49] who studied the influence of container size on root development and on the growth, yield and quality of various fruit crops. Phala et al. [50] also found that the number of inflorescences increased due to size of different growing media bags, since container size has impacts on water-holding capacity of growing media, which has been reviewed by Passioura [51] and Poorter et al. [47].

Container type was also evaluated by some authors such as Manole et al. [49] who found that the type of pot affects the biological activity and growth parameters of the plants. This finding is confirmed by Sharma and Godara [52] who reported that in case of PVC pots, due to negligible evaporative cooling results in the higher temperature of the rooting media, which ultimately shows better root-shoot growth of the plant in the winter month. Sharma et al. [41] suggested that the cultivation of strawberries can be undertaken as a commercial cultivation by utilizing a PVC pot filled with cocopeat, perlite and vermicompost media in a ratio of 3:1:1. And these interaction PVC pot substrates showed a good performance of growth, yield and quality of strawberry cv. Chandler. Many studies have been conducted to optimize the pot volume and substrate for growing strawberries in pots, and most of the researches reported that the higher the substrate volume is, the greater the vegetative growth and fruit yield [47, 53]. Godara et al. [48] found that plants grown in larger volume pots provided a greater quantity of fruits compared to smaller pots. In addition, the pot volume had effects on the timing of peak harvest, which is demonstrated by some authors like Massetani et al. [53] who evaluated the growth of "Elsanta" strawberry on certain pot volumes such as 250-cm^3 pots and 150-cm^3 pots and revealed that the earliest floral differentiation was observed on strawberries produced on larger pots (250 cm^3). Therefore, in pot culture, establishing an appropriate range of substrate volumes considering the growing period is crucial as concluded by Lee et al. [54]. For example, according to Lee et al. [54], a pot volume of 830 cm^3 and coir are most suitable for about 6 months of strawberry pot cultivation [54]. A small pot implies a small volume of substrate, which reduces the availability of resources for plants, hindering root development because when the root system is constrained to a small rooting volume, the equilibrium between roots and shoots can be disturbed, which can have a detrimental impact on plant development [47]. In addition, smaller pots more rapidly due to their higher surface area-to-volume ratio. Thus, optimizing the appropriate pot volume and substrate using pot cultivation can enhance the efficiency of the cultivation area, thereby increasing the yield. To conclude, in strawberry pot culture, proper management of appropriate root volume, irrigation volume and drainage rate is essential depending on the characteristics of the substrates (physical and chemical properties) and the growth characteristics of the plant [55].

2.2 Irrigation and nutrient management of strawberry growing in substrate

Irrigation management during the strawberry growing season is of great importance for fruit yield and quality. Reducing evaporation and leaching losses with maximum yield is the primary objective of any soilless strawberry grower in order to maximize water use efficiency [56]. For this reason, strawberry growers are using bagged substrate systems which have shown the highest early and total yield per m^2, energy efficiency, water efficiency as well as impact on economic return [57].

Treatments										
mmol/L	NH_4^+	K^+	Ca^{2+}	Mg^{2+}	NO_3^-	Cl	SO_4^{2-}	$H_2PO_4^-$	H^+	Si
	0,50	4,00	3,50	1,25	10,00	0,00	1,25	1,00	0,00	0,00
μmol/L	Fe	Mn	Zn	B	Cu	Mo				
	20,00	15,00	7,00	10,00	0,50	0,50				

Disclaimer this example is just used to identify molar concentration. The use of this balance must be analyzed as guide, and every fertigation recipe must adapt to growing regions.

Table 1.
Concentrations of different elements (macro- and micronutrients) in the nutrient solution used in strawberry.

Nutrient management in the fertigation is also an important strategy to optimize fruit quality and shelf life for strawberries [58]. For example, in the strawberry semi-hydroponic system one of the most critical points is the nitrogen management by fertigation [59]. Nitrogen availability affects strawberries productivity and fruit quality [60, 61]. As such, higher total nitrogen content in the substrate leads to better nitrogen utilization efficiency and increases the nitrogen absorption of strawberries [62]. Both nitrogen forms, nitrate and ammonium salts, are absorbed and metabolized by plants [63] and are playing important metabolic functions, such as growth, development and production, and affect fruits appearance, durability and flavor [64].

When composing the fertilizer recipe, EC of the water must be taken into account. Changing irrigation EC levels not only changes the total amount of fertilizer required but it also changes the required ratio between the nutrients in the fertilizer. For strawberries, EC is kept constant at 2 dS m^{-1}. A study by Sarooshi and Cresswell [65] showed that reducing EC from 3 to 2 dS m^{-1} at the beginning of fruit set increased berry size compared to a constant EC of 2 dS m^{-1}, while decreasing yield when increasing from 2 to 4 dS m^{-1}. An experiment with tomato also revealed that marketable yield decreased by about 5% for each dS m^{-1} above 2 dS m^{-1} [66]. **Table 1** shows an example of a strawberry recipe used in Hortitool Consulting.

Experience from the field shows that conducting fertigation with high EC leads to salt accumulation in substrate that will require action leaching it frequently, through the growing season, also as strategy. This has to be included because once growers use the same substrate multiple years, as seen in south of Europe or Mexico, it is a must to maintain the salt accumulation as low as possible.

3. Growing blueberries in substrate

Blueberry plants grow in acidic soils showing an optimal growth at pH between 4.0 and 5.5 [14] and have been classified as calcifuge plants [67]. Specifically, highbush blueberries require a soil pH between 4.5 and 5.5, while rabbiteye blueberries require a soil pH of around 4.2 to 5.0 [68]. These soil types are not common in all areas of the world where blueberries are grown, and for this reason, blueberry growers amend soils with pine bark and additional sulfur [69]. Blueberry has long been described as a plant with low nutritional requirements and a slower growth rate [70]; therefore, the physiological properties of blueberries determine their potential application in soilless culture. Soilless cultivation, using substrates to replace soil, is an optional way to plant blueberries [71]. The use of substrates can effectively avoid cultivation constraints such as soil-borne diseases and insect pests, and also control production, which can ensure high yield and high quality of the fruits [72].

Many studies have revealed that blueberries exhibit high productivity from the first year of cultivation on substrate [11, 69], as blueberry plants grown on substrates have vigorous and early growth. Some substrates have proven effective due to their composition of materials such as coco fiber, perlite, turf, bark, wood fiber, peat moss and sand [14, 73]. An industry standard growing substrate for containerized blueberry production has not yet been identified [74]. However, several studies have proven that coco fiber is a very common component of substrates used in blueberry production [14] due to its favorable physical and chemical characteristics [69, 75]. According to Ortiz-Delvasto et al. [12], the blueberry cultivar Legacy has better growth and yield as well as a larger fruit diameter in pure coco fiber compared to a mixed substrate of coco fiber and peat. Besides, substrate standards have been set by private substrate producers based on longevity and chemical and physical stabilities. For example, a mixture of substrates such as peat moss, coco peat/coir and perlite is used by blueberry growers in a ratio of 40.40.20 as an average standard.

Determining the physical and chemical properties of substrates is essential for their effective use, as they determine the productive potential of plants, providing an environment in which roots will develop, which directly influences plant growth and development [76]. Due to the susceptibility of blueberry fine roots to root rot in poorly drained soils, the growing medium must provide adequate physical properties and drainage to avoid any risk of disease [77]. For example, for blueberry nursery cultivation, a typical growing medium includes a total porosity of 50–85%, with appropriate levels for potted blueberries at the higher end of this range (76–86%) [14]. Appropriate chemical properties of the substrate must also be maintained during potted blueberry production, including pH (4.2–5.5) and electrical conductivity (< 2.0 mS•cm^{-1}) [14, 78].

The composition of the substrate can affect the native pH of the substrate; for example, the pH is the lowest in substrates composed of pine bark, the highest in substrates composed of coconut fiber and intermediate in substrates composed of sphagnum peat moss. In addition, the pH of the substrate can change rapidly depending on the pH of the irrigation water or fertilization solution and the uptake of nutrients by plants, and in general, substrates do not have a high pH buffering capacity; therefore, the concentrations of carbonate and bicarbonate in the water (collectively called alkalinity) must be taken into account when managing the pH of the substrate [69].

Regarding electrical conductivity, when the electrical conductivity (EC) value is higher than 1.0 dS/m, the growth, production and quality of blueberries are affected. In hydroponic systems, a nutrient solution with an electrical conductivity between 0.5 and 1.0 dS/m can be used without affecting the yield and quality of the fruits, noting as an example the blueberry cv. Biloxi that has been shown to be sensitive to substrate salinity as reported by Frías-Ortega et al. [79].

3.1 Container effect on blueberry cultivation in substrates

Historically, blueberries have been propagated and grown in containers in nurseries, but only for a short period of time before planting. In recent years, interest in containerized blueberry production has grown as an alternative to growing blueberries in the soil [80–82]. Container-grown plants typically have limited space for root growth, unlike those grown in traditional field environments, but blueberry bushes are known for their shallow root systems, so the root systems of many 2- to 4-year-old blueberry cultivars were concentrated primarily in the top 20 cm of soil, making them suitable for growing in pots [83, 84]. For container blueberry growers, container size and shape are important decisions. Reduced root volume can affect plant growth as it negatively affects root respiration, nutrient uptake, photosynthesis and subsequently flowering and biomass accumulation and distribution [47]. As container size increases, shoot and root biomass increases [85]. For example, in central Florida, container sizes commonly used in commercial blueberry production range from 56 to 95 L, which allows for high yields in the first year, ranging from 0.9 kg/plant to over 2 kg/plant depending also on the cultivar and fertilizer rates used [80, 84, 86]. Container sizes of 50.5 and 40.4 L have also been evaluated for blueberry cultivation and results during the first 2 years after planting have proven that this size range is adequate for root growth and plant parameters [87]. Smaller containers, less than 38 L, can have a negative effect on yields especially during extreme weather events, for example, in case of strong winds, plants in small containers can be knocked over, requiring trellising or other anchoring mechanisms [69, 84, 88]. Container parameters also have an impact on production management, including irrigation and fertilization. According to Passioura [51], container height has a positive effect on drainage water content; however, very short containers can lead to a higher risk of hypoxia for plant roots [51]. The height and width of pots direct the distribution of water between the surface and the bottom of the container [89] and unequal water distribution within the container can lead to uneven distribution of root systems [47] affecting both water-holding capacity and aeration in the substrate. Generally, blueberry growers prefer wider containers over taller ones [90]. Containers with many holes or grids for drainage are also recommended to prevent root diseases such as *Phytophthora* root rot and to help manage substrate salinity [69].

Regarding container shape, there is no consensus on the effect that round or square shape or color (black or white) would have on blueberry root growth. Finally, container blueberry production has many unknowns; therefore, recommendations should be made to container blueberry growers on container size, substrate composition and cultivar choice. There are cultivars very suitable for this cultivation system (pots under tunnels) such as Bluetta cv showing an advanced harvest period of about 3 weeks [13, 91]; thereby, the development of cultivars particularly suitable for blueberry cultivation in containers in different climates around the world should be the subject of future research on this blueberry cultivation system to improve it [92].

3.2 Irrigation and nutrient management of blueberry growing in substrate

3.2.1 Irrigation

Management, particularly irrigation and fertilization, is essential to achieve the high plant vigor that characterizes blueberry production on substrate [69]. Substrates were introduced for container crop production with the aim of increasing drainage by maintaining a relatively high proportion of air-filled porosity compared to traditionally used mineral soils [93]. This has led to a practice of applying excess water to eliminate any risk of underwatering [94]. However, excessive watering leads to inefficient use of water resources and subsequent leaching or runoff of applied agrochemicals [95]. Overwatering of pots also has a significant impact on root system development and can lead to plant collapse. For these reasons, container crop producers need to be more conscientious in their water use [96]. Various efforts to reduce water loading in container production, for example, in the case of ornamental crops, include different irrigation schemes [97], modeling of crop water use [98], varying crop spacing and arrangement [99], and sensor-controlled irrigation [100], although there is no clear alternative to overhead irrigation in container nursery production [96]. But only innovators or early adopters have begun to take significant steps to implement water-saving practices.

In Florida, USA, most potted blueberry growers use two or more emitters per pot with the goal of maintaining drainage at 10% of total water input to achieve uniform substrate moisture. Substrate moisture sensors have also been used as an alternative, less labor-intensive irrigation approach. Drip irrigation, although expensive, remains the most widely used system due to better water control and greater distribution uniformity [69].

Irrigation based on leachate fraction is also a common practice in blueberry cultivation on substrate. This is the proportion of leachate volume to the volume of irrigation/fertigation solution applied. Previous research has targeted leachate fractions between 15% and 25%, while those above 25% are not recommended because they cause excessive nutrient losses and saturated conditions can promote *Phytophthora* or *Pythium* root diseases [69]. Moreover, the solution that flows from the containers provides a good representation of the chemical conditions of the substrate. Therefore, periodic collection and analysis of leachate is essential to monitor the pH and electrical conductivity (EC) of the substrate.

When water salinity is high or heavy fertigation regimes are used, salinity buildup in the substrate can be a problem and the ability to cycle between irrigation and fertigation events is useful to percolate salts out of containers if necessary. Irrigation water may contain carbonates and bicarbonates, which characterize most blueberry-growing areas [101], and to neutralize the alkalinity, acidification of the water must be achieved by sulfuric acid injection or sulfur dioxide generators [102–104].

The water status of blueberry should be monitored when in this culture system. The level of tissue dehydration in blueberry can be estimated by a direct method, which is the measurement of relative water content (RWC) [12, 105]. In general, RWC values fluctuate between 98% (turgid leaves) and 40% (severely dehydrated leaves) [105]. Blueberry has a shallow fibrous root system that is unable to provide sufficient water lost through leaf transpiration under abiotic stress conditions [106], resulting in a higher wilting point (up to 70% depending on cultivars), compared to other species [12]. Part of the fruit production could be attributed to the greater water-holding and ion-exchange capacity, which allows adequate nutrient availability compared to field conditions [107].

3.2.2 Nutrient management

Blueberries have lower nutrient requirements than most crops and thrive in acidic soils with limited availability of essential nutrients such as nitrate nitrogen (NO_3-N), phosphorus (P), potassium (K), calcium (Ca) and magnesium (Mg) [108]. Nitrogen is the primary nutrient applied to blueberries and is required annually. Blueberries primarily acquire the ammoniacal (NH_4) form of N, rather than NO_3-N unlike most crops, due to low nitrate reductase activity in roots and leaves [109, 110]. All nutrients that apply to blueberries include P, K, Ca, Mg, elemental sulfur S (which is used to lower soil pH), iron (Fe), boron (B), copper (Cu) and zinc (Zn) [111]. In the substrate, all of these elements must be applied by fertigation to obtain a balanced recipe. **Table 2** presents a common recipe used to grow southern highbush blueberries (SHB) studied by Hortitool Consulting in 2019.

Treatments										
mmol/L	NH_4^+	K^+	Ca^{2+}	Mg^{2+}	NO_3^-	Cl	SO_4^{2-}	$H_2PO_4^-$	H^+	Si
	3,00	1,50	2,00	1,00	4,50	0.00	2.50	0.00	0.00	0.00
µmol/L	Fe	Mn	Zn	B	Cu	Mo				
	20,00	10,00	7,00	10,00	0,50	0,50				

Disclaimer this example is just used to identify molar concentration. The use of this balance must be analyzed as guide and every fertigation recipe must adapt to growing regions.

Table 2.
Concentrations of different element (macro- and micronutrients) in the nutrient solution used in southern highbush blueberries (SHB): vegetative stage use in substrate mix with Baltic peat, coco pith and perlite.

For blueberries grown in substrate, all essential nutrients must be provided in the fertilizer solution, as growing media do not contain essential nutrients and only retain positively charged nutrients (called cations). Therefore, for successful production of potted blueberry plants, developing nutrient management strategies requires understanding the appropriate amount of nutrients that blueberry plants require at different growth stages. Leachate analysis plays a crucial role in monitoring nutrient sufficiency [69] and controlling nutrient losses; hence, reducing these losses is an important issue for this crop [82]. Leaching can increase significantly if inappropriate water and nutrient management practices are used [112]. In some cases, plants absorb more water than nutrients, which can lead to salinity accumulation. Therefore, in cases of high salinity, substrates should be flushed with large volumes of acidified water to remove excess salts. The timing and method of fertilizer application allows growers to influence the nutrient status of the root zone [113] as well as maintain proper pH levels in the root zone that will promote healthy plant growth [74]. Keeping target EC and pH levels in mind is important for container blueberry growers, allowing for proper nutrient levels to be provided at the right time to manage substrate fertility for healthy blueberry growth and higher fruit yield and quality.

Currently, there are several different strategies available for blueberries and the fertilizer regime and rate should be adjusted based on plant growth stages based on commercial practices [87]. Optimal fertilization of potted northern highbush blueberry *Vaccinium corymbosum* "Duke" was studied for different production methods (organic and conventional) by Clark and Zheng [74] who found that high fertilizer application rates are unnecessary and can be detrimental to plant growth and fruit production. According to Kingston et al. [14], fertilizer should be provided by continuous fertigation, which would avoid the risk of reduced nutrient uptake. Nutrient deficiencies can be determined by foliar nutrient analysis based on plant phenology [69]. For example, for northern highbush blueberry, leaf tissue analysis is a useful tool to monitor nutrient requirements [102]. It is also recommended to introduce into the blueberry fertigation strategy also foliar fertigation to integrate nutrient deficiencies, especially to correct calcium deficiency on fruits, as reported by Gerbrandt et al. [114].

4. Growing raspberry in substrate

Various types of substrates are used for raspberry cultivation including peat substrate, perlite, coke, mineral wool and coconut fiber. The composition of the substrate influences the yield of raspberries and any change in the composition has an impact on the growth and development of the plants as well as on the biological properties of the raspberries produced [115]. Peat has been an important component for this crop for a long time because of their excellent water-holding capacity and good aeration [116], while findings carried out by Carlen et al. [116] demonstrated that substrates with native organic components function very well without peat. Peat has long been an important component of this crop due to its excellent water-holding capacity and good aeration [116]. Some authors have demonstrated that substrates containing native organic components can replace peat and even work very well [116]. In recent years, coconut fiber, biochar and coarse bark have been proposed to replace peat on the market [93]. These materials can also be mixed with peat to be better suited for raspberry production, as revealed by Qiu et al. [18]. In addition, greenhouse raspberry cultivation in pots containing coconut fiber substrate is proven to be greater and even typical by some authors [108].

Just like when changing fertilization with different nutrients, any change in the composition of the substrate can change the composition and content of bioactive compounds in raspberries. There are different organic additives that can be mixed into the substrate such as sheep wool, hemp fibers, a concentrated organic component obtained by steam treatment of fragmented wood materials and sunflower seed biochar and that have an influence on the total polyphenol content of the fruits as reported by some studies. Coconut substrate with organic additives significantly increases the polyphenol content of raspberries. For example, according to the results of Balawejder et al. [115], coconut substrate enriched with sheep wool increases the polyphenol content of fruits by about 37.7% compared to fruits grown on a substrate containing 100% coconut fibers. These additive components can also significantly change the enzymatic activity of raspberries, which show the low level of stress and good condition of the plants.

4.1 Container effect on raspberry cultivation in substrates

Growers in several regions grow raspberries in containers to avoid soil problems for greater flexibility in harvest times. However, container production is more expensive than soil production. In this case, raspberry growers try to have a high yield with high-quality fruits to compensate for the costs [110]. Root conditions are among the main factors that can influence the profitability of raspberry crops on substrates. Therefore, different pot sizes, pot types and pot colors have been evaluated in several researches for their influence on the root surface of raspberries, which have a direct effect on yield, fruit size and fruit quality [116]. Larger pots have an extremely positive influence on yield and fruit size, but berry quality is not influenced when raspberry is grown in smaller pots, as revealed by Carlen et al. [116] when evaluating the production of "Tulameen" (floricane) raspberries in different pot sizes. Furthermore, in recent years, several agroscope trials have been conducted on floricane raspberries and have recommended suitable pot sizes for better yield and fruit size, such as 10-L pots with two plants pot^{-1} and three pots m^{-1}. The type of drainage system in the pots (pots with a grid and small 0.03-m heels at the bottom or pots with holes directly on the bottom) has also been studied by some authors such as Carlen et al. [116] and Linnemannstöns [17] who demonstrated that this parameter is important, as containers must allow good drainage to increase raspberry yield and fruit weight, which increases with pots with a grid and small heels on the ground. An example is strawberry production, which shows increased yield and fruit weight when grown in pots with a grid and small heels on the ground. This positive effect is probably due to better air circulation and drainage of the substrate and thus lowers salt stress [116]. A large number of holes in the bottom and a rim at the bottom of the container are also necessary, so that the pots do not rest directly on the ground. In addition, a slight elevation of the soil surface is necessary to ensure that drainage water can drain well. Otherwise, this can quickly lead to overwatering and plant loss [17]. Container shape (round or square) is another parameter studied for raspberry production but has no influence on fruit yield, size and quality as reported by some authors. However, pot color is another characteristic that, interestingly, has been shown to influence early ripening and yield of raspberries. For example, according to Carlen et al. [116], harvest started about 20 days later and fruit yields were slightly higher for raspberry plants in white pots than in black pots [116].

4.2 Irrigation and nutrient management of raspberry growing in substrate

4.2.1 Irrigation of raspberry growing in pots

Growers use pressure-compensating emitters for the irrigation of raspberries, which will help ensure that each plant receives the same amount of water, and helps prevent overwatering or under-watering [117]. Typically, 30–50% of applied water should drain from the pots, according to pot growers, to ensure adequate leaching and avoid salt accumulation [110]. In our practices, we use a percentage of water for drainage ranging from 20 to 25% as standard percentage for plant growth, depending on the growth stage of the plant, water quality and salt concentration. Also, an optimal pH ranges from 6.0 to 6.5 and an optimal EC < 30 ppm, sodium <50 ppm and chlorine with an EC < 3 mS/cm. Maximizing yields with minimal irrigation inputs by increasing plant water use efficiency becomes essential [118]. In this context, a study carried out in semi-arid climate by Lepaja et al. [119] using a technique of partial rootzone drying (PRD) on raspberries "Glen Ample" and "Glen Prosen" found that PRD is an ideal water-saving technique that can improve the physical and chemical qualities of fruits. This technique involves keeping about half of the root system adequately watered, while the other part is allowed to dry out. To achieve this, both halves are watered alternately from time to time [120]. This process is explained in several species by the fact that when part of the root system dries out, a root-shoot signaling mechanism is triggered; these signaling molecules are transported *via* the xylem to the leaves, where they result in partial stomatal closure [121].

In some cases, PRD can reduce stomatal conductance, thus reducing water loss [122]. This technique does not affect yield in raspberries, and differences in the volume of the water supplied resulted in some reduction of vegetative growth of raspberry canes as demonstrated by Lepaja et al. [119]. PRD is an advantage under semi-arid climate conditions where raspberry cannot be grown without irrigation, also it can be successfully applied to raspberry orchards to reduce vegetative growth and increase total yield.

4.2.2 Nutrient requirements

Nutrient requirements of raspberries depend on crop size, habitat, plant age, soil type, irrigation and cultivar [123]. Raspberries rapidly absorb large amounts of nutrients during vegetative growth [17] and mineral nutrient requirements change during different stages of plant development [124]. Fertilization is typically based on

Treatments										
mmol/L	NH_4^+	K^+	Ca^{2+}	Mg^{2+}	NO_3^-	Cl	SO_4^{2-}	$H_2PO_4^-$	H^+	Si
	1.00	3,00	4,50	1,75	11,50	0,00	2,00	1,00	0,00	0,00
µmol/L	Fe	Mn	Zn	B	Cu	Mo				
	20,00	15,00	7,00	10,00	0,50	0,50				

Disclaimer this example is just used to identify molar concentration. The use of this balance must be analyzed as guide and every fertigation recipe must adapt to growing regions.

Table 3.
Concentrations of different element (macro- and micronutrients) in the nutrient solution used in raspberry: vegetative stage use in coir substrate crop.

soil tests every few years and observations of annual growth (cane number, diameter and height, and lateral length of fruiting bodies), yield, leaf color and fruit quality [125]. However, container raspberry growers, to determine actual nutrient levels in raspberries, adopt leaf tissue analysis, which is essential for assessing plant nutritional status [117]. Commercial raspberry growers develop fertilization programs based on initial nitrogen recommendations, with adjustments to nitrogen and other nutrients as needed [125], and the highest nitrogen demand occurs during the period of intense growth, flowering and fruit development [125, 126]. During harvest, reduced leaf growth results in less fertilizer requirement, and if fertilization is too high, fruit becomes too soft [17]. For example, raspberry growers in the northern hemisphere typically make their decisions about nitrogen fertilization of current-year canes between late July and early August. In intensive raspberry plantations in the continental European climate, fertilization typically begins at the flower bud stage in June and continues until mid-harvest in August, using special water-soluble fertilizers containing P and K as well as various micronutrients [111, 127]. For container-grown raspberries, due to the low soil volume and watering frequency, fertilization is applied *via* the irrigation system, which is a convenient method to provide nutrients such as nitrogen to raspberries [117]. Besides, commercial soluble fertilizers that contain a complete nutrient package are now available for these growers and most of these formulations have been developed for both flower and foliage crops. In this context, **Table 3** shows an example of a nutrient recipe to be used at the vegetative stage in a coco coir substrate culture adopted by Hortitool Consulting since 2019.

5. Conclusion

The world of berries and its techniques are vast and somehow complex. Specific varieties and photoperiod (i.e., strawberry short-day vs. ever bearers), plant material (i.e., plugs vs. trays), planting dates, substrate types, climate and moreover market-targeted cropping are some of main points to address in production strategies. In this review, it is visible that research has been on early stage of having specifics guidelines, especially in blueberries, where anyone could use it and apply in every condition and if it is applicable to day-to-day growing. Some other researches have contradictive results and different from what in the farming is used, as an example the EC target in strawberries where EC on farming level is far lower than that used for research purpose. In other hand, for blueberries in substrate, trials studies need to be closer to farm conditions in order to learn the different type of substrates that are used commercially and more researches are need to be done on fertigation nutrient requirements, once the industry has been much more developed in the Southern Highbush genetics, mainly in the last 20 years, and little has been done on this matter. On the Rubus side, more works are done on raspberries than on blackberries, in research from US and UK universities, and extension acts in governmental institutes. Nowadays, most of the substrate selection is a substrate company recommendations bringing the trends, and the fertigation programs on the hand of private consulting companies and university research. With that said, more close relations must be created between research and real farming operations from the private sector, growers, substrate, fertilizer companies and consulting companies to be able to determine the best and optimal practices to adapt and customize in each crop and growing region.

Moreover, this summary of these techniques raises the need to gather technology adaptation on the farm level to address the needs to confirm the ideal substrate mixes, more specific fertigation for each growing areas, species and plant types, in which climate (light levels, day length and other factors) can be different. All of these limitations will require an industry effort for all the learning gathered be confronted and create better growing protocols for stakeholders to explore better the potential of the genetics adapted to their own business plans.

Acknowledgements

I want to thank my colleagues Marco and Haifa for the contributions to this chapter, with special thanks for the hard job of review over and over this data and be so collaborative in this participation. Furthermore, I would like to address the hours spend on the field with our great growers and clients where we learn and teach the crops we fell in love with.

Author details

Haifa Ben Hadj Daoud[1], Marco Rosario Butera[2] and Jorge Pedro Rosa Duarte[3*]

1 Department of Biological Sciences and Plant Protection, High Institute of Agronomy of Chott Mariem-University of Sousse, Sousse, Tunisia

2 Mrb-Agriconsulting, San Bonifacio (Vr), Italy

3 Hortitool Consulting Lda, Faro, Portugal

*Address all correspondence to: jdhortitool@gmail.com

References

[1] Intelligence SO. Berries in the World Report. Finland: SITRA; 2008

[2] Hera O, Teodorescu R, Sturzeanu M. Blueberry (*Vaccinium Corymbosum*) breeding programme in the main cultivating countries. Scientific Papers Series B Horticulture. 2021;**LXV**:82-89

[3] Raja WH, Kumawat KL, Sharma OC, Sharma A, Mir JI, Nabi SU, et al. Effect of different substrates on growth and quality of strawberry cv. Chandler in soilless culture. The Pharma Innovation Journal. 2018;**7**(12):449-453

[4] Sahoo S, Sahoo D, Sahoo KK. Optimization of an efficient hydroponic cultivation method for high yield of strawberry plants. South African Journal of Botany. 2024;**167**:429-440

[5] Unal N, Kahramanoglu I, Hajizadeh HS, Bilgin J, Volkan O. Effect of different growing media on pomological and phytochemical parameters of Fragaria vesca 'Yellow Wonder' and Fragaria ×ananassa 'Camarosa': A comparative study. Plant Biology. 2023;**23**:400

[6] British Summer Fruits. The impact of Brexit on the UK soft fruit industry. British Summer Fruits Seasonal Labour Report. 2017;**25**:280-286. DOI: 10.1080/09548963.2016.1241368

[7] Wanas AL, Mennat Allah IK. Effect of some soilless culture systems on growth and productivity of strawberry plants. International Journal of Agricultural Sciences and Technology. 2022;**2**(1):18-29

[8] Hernández-Martínez NR, Blanchard C, Wells D, Salazar-Gutiérrez MR. Current state and future perspectives of commercial strawberry production: A review. Scientia Horticulturae. 2023;**312**(15):111893

[9] Osorio R, Cáceres C, Covarrubias JI. Vegetative and physiological responses of 'emerald' blueberry to ammonium sources with nitrification inhibitor. Journal of Soil Science and Plant Nutrition. 2020;**20**:1-9

[10] Mateljan G. World's Healthiest Foods, 2nd Edition: The Force for Change to Health- Promoting Foods and New Nutrient-Rich Cooking. 2nd ed. Glendale, United States, California: George Mateljan Foundation; 2015. Available from: http://www. whfoods.com/genpage.php?tname=foodspice&dbid=8

[11] Fang Y, Nunez GH, Silva MND, Phillips DA, Munoz PR. A review for southern highbush blueberry alternative production systems. Agronomy. 2020;**10**:1531

[12] Ortiz-Delvasto N, Garcia-Ibañez P, Olmos-Ruiz R, Bárzana G, Carvajal M. Substrate composition affects growth and physiological parameters of blueberry. Scientia Horticulturae. 2023;**308**:111528

[13] Pinto RM, Mota M, Oliveira CM, Oliveira PB. Effect of substrate type and pot size on blueberry growth and yield: First year results. Acta Horticulturae. 2017;**1180**:ISHS 2017

[14] Kingston PH, Scagel CF, Bryla DR. Suitability of sphagnum moss, coir, and Douglas fir bark as soilless substrates for container production of highbush blueberry. HortScience. 2017;**52**:1692-1699

[15] Durán-Soria S, Pott DM, Will F, Mesa-Marín J, Lewandowski M, Celejewska K, et al. Exploring genotype-by-environment interactions of chemical composition of raspberry by using a metabolomics approach. Metabolites. 2021;**11**:490

[16] Kalušević A, Salević A, Đorđević R, Veljović M, Nedović V. Raspberry and blackberry pomaces as potential sources of bioactive compounds. Ukrainian Food Journal – Food Technologies. 2016;**5**(3):485-491

[17] Linnemannstöns L. Substrate cultivation of raspberry in Germany. Acta Horticulturae. 2020;**1277**:ISHS. DOI: 10.17660/ActaHortic.2020.1277.24

[18] Qiu C, Gaudreau L, Nemati R, Gosselin A, Desjardins Y. Primocane red raspberry response to fertigation EC, types of substrate and propagation methods. European Journal of Horticultural Science. 2017;**82**(2):72-80

[19] INIA. El cultivo de la fresa. Conafrut. Boletín técnico Nº 12. Vol. 28. Lima, Perú: INIA; 1997. p. 5

[20] Durner EF, Poling EB, Maas JL. Recent advances in strawberry plug transplant technology. American Society for Horticultural Science. 2002;**12**:545-550

[21] Thakur M, Shylla B. Effect of different soilless substrates on flowering, yield and fruit quality of strawberry (*Fragaria x ananassa* Duch.) cv. Chandler under protected conditions. International Journal of Current Microbiology and Applied Sciences. 2018;**7**(4):2830-2836

[22] Walters KJ, Behe BK, Currey CJ, Lopez RG. Historical, current, and future perspectives for controlled environment hydroponic food crop production in the United States. HortScience. 2020;**55**:758-767

[23] De-Rijck G, Schrevens E. Distribution of nutrients and water in rockwool slabs. Scientia Horticulturae. 1998;**72**:277-285

[24] Bhat R, Geppert J, Funken E, Stamminger R. Consumers perceptions and preference for strawberries—A Case study from Germany. International Journal of Fruit Science. 2015;**15**(4):405-424

[25] Nihad GA, Malik GA, Jamal YA, Yahia AO, Rolston SH. Composition of soilless substrates affect the physiology and fruit quality of two strawberry (*Fragaria × ananassa* Duch.) cultivars. Journal of Plant Nutrition. 2018;**41**(18):2356-2364

[26] Cantliffe DJ, Castellanos JZ, Paranjpe AV. Yield and quality of greenhouse-grown strawberries as affected by nitrogen level in coco coir and pine bark media. Proceeding of the Florida State Horticultural Society. 2007;**120**(1):157-161

[27] Hochmuth G. Containerized strawberry transplants reduce establishment period water use and enhance growth and flowering compared with bare-root plants. Hort Technology. 2008;**16**:46-54

[28] Anagnostou K, Vasilakakis MD, Gerasopoulos D, Olympois CH, Passam H. Effects of substrate and cultivar on earliness, plant productivity, and fruit quality of strawberry. Acta Horticulturae. 1995;**379**:267-274

[29] El-Sayed SHF, Abul-Soud HA, Dam M. Effect of substrate mixtures and nutrient solution sources on strawberry plants under closed hydroponic system. International Journal of Product Development. 2016;**21**:97-127

[30] Ayesha R, Fatima N, Ruqayya M, Faheem H, Qureshi KM, Hafiz IA, et al. Influence of different growth media on the fruit quality and reproductive growth parameters of strawberry (*Fragaria* × *ananassa* Duch.). Journal of Medicinal Plants Research. 2011;**5**:6224-6232

[31] Linardakis DK, Manios BI. Hydroponic culture of strawberries in plastic greenhouse in a vertical system. Acta Horticulturae. 1991;**287**:317-326

[32] Paraskevopoulou G, Grefiadellis M, Paroussis E, Maloupa E, Gerasopoulos D. Precocity, plant productivity and fruit quality of strawberry plants grown in soil and soilless culture. Acta Horticulturae. 1995;**408**:109-117

[33] Lopez-Medina J, Perablo A, Flores F. Closed soilless system: A sustainable solution to strawberry crop in Huelva Spain. Acta Horticulturae. 2004;**649**:213-215

[34] Wang D, Gabriel MZ, Legard D, Sjulin T. Characteristics of growing media mixes and application for open-field production of strawberry (*Fragaria ananassa*). Scientia Horticulturae. 2016;**198**:294-303

[35] Maher MM, Shylla B, Sharma DD, Sharma U. Influence of different soilless substrates and jeevamrit on flowering and fruiting behaviour of strawberry (*Fragaria* × *ananassa* Duch.) cv. Chandler. Journal of Pharmacognosy and Phytochemistry. 2020;**9**(4):428-432

[36] García-López DA, Cruz-Ortega R. Evaluation effects of alternative substrates for soilless cultivation of strawberry (*Fragaria* × *ananassa*). Nexo Revista Científica. 2023;**36**(6):831-838

[37] Bartczak M, Pietrowska M, Knaflewski M. Effect of substrate on vegetative quality of strawberry plants (*Fragaria* × *ananassa* Duch.) produced by a soilless method. Folia Horticulturae Ann. 2007;**19**(2):39-46

[38] Rivera-del Rio R, Pineda-Pineda J, Avitia-Garcia E, Castillo-González AM, Vargas-Hernandez M. Alteration of physical properties of substrates and accumulation of nutrients in strawberry hydroponic systems (*Fragaria x ananassa* Duch.). Acta Horticulturae. 2017;**1170**:679-686

[39] Jafarnia S, Khosrowshahi S, Hatamzadeh A, Tehranifar A. Effect of substrate and variety on some important quality and quantity characteristics of strawberry production in vertical hydroponics system. Advances in Environmental Biology. 2010;**4**:360-363

[40] Ameri A, Tehranifar A, Davarynejad GH, Shoor M. The effect of substrate and cultivar in quality of strawberry. Journal of Biological and Environmental Sciences. 2012;**6**(17):181-188

[41] Sharma R, Bakshi P, Kour K, Jasrotia A, Kumar R, Maanik Abrol V, et al. Effect of container size and growing media on growth, yield and quality of strawberry (*Fragaria* × *ananassa* Duch.). Agricultural Mechanization in Asia. 2022;**53**:09. ISSN: 00845841

[42] Key JL. Hormone and nucleic acid metabolism. Annual Review of Plant Physiology. 1969;**20**:449-474

[43] Ulrich D, Olbricht K. A search for the ideal flavor of strawberry – comparison of consumer acceptance and metabolite patterns in *Fragaria* × *ananassa* Duch. Journal of Applied Botany and Food Quality. 2016;**89**:223-234

[44] Sharma VK. Effect of soilless growing substrates on vegetative growth

and roots of strawberry cv. sweet Charlie. The Ecoscan. 2015;**9**:89-91

[45] Davis JN. Effect of nitrogen, phosphorous and potassium fertilizers on the non-volatile organic acids of tomato fruit. Journal of the Science of Food Agriculture. 1964;**15**:665-673

[46] Hassan AH, Khereba AH, El-Kattan MH, Noha G, ElRahman A. Effect of various organic substrate culture and container types on productivity and fruit quality of strawberry (*Fragaria* × *ananassa*) cv. festival. Research Journal of Agriculture and Biological Sciences. 2011;7:379-387

[47] Poorter H, Bühler J, Van Dusschoten D, Climent J, Postma JA. Pot size matters: A meta-analysis of the effects of rooting volume on plant growth. Functional Plant Biology. 2012;**39**:839-850

[48] Godara AK, Sharma VK, Beniwal V. Effects of substrates and containers on root system of hydroponically grown strawberry (*fragaria* × *ananassa* duch.) under greenhouse. Bangladesh Journal of Botany. 2016;**45**(5):1083-1089

[49] Manole MS, Rosu M, Dobrin E, Ciofu R, Gheorghita N. The influence of pot type on the biological activity of the substrate and the seedlings growth. Analele Universitatii din Craiova. 2008;**13**:313-320

[50] Phala MG, Campo JL, Oliveira PB. Optimizing a high-density soilless culture system for strawberry production. Acta Horticulturae. 2012;**926**:503-507

[51] Passioura JB. Viewpoint: The perils of pot experiments. Functional Plant Biology. 2006;**33**:1075-1079

[52] Sharma VK, Godara AK. Response in strawberry (*Fragaria* × *ananassa* Duch. 'Sweet Charlie') growth to different substrates and containers under greenhouse. International Journal of Current Microbiology and Allied Sciences. 2017;**6**(11):2556-2568

[53] Massetani F, Savini G, Neri D. Effect of substrate and container type in the strawberry soilless cultivation. Acta Horticulturae. 2017;**1156**:295-300

[54] Lee H, Cui M, Lee B, Hwang H, Chun C. Optimization of the pot volume and substrate for strawberry cultivation in a hydroponic system. Horticultural Science and Technology. 2023;**41**:634-644

[55] Pignata G, Casale M, Nicola S. Water and nutrient supply in horticultural crops grown in soilless culture: Resource efficiency in dynamic and intensive systems. In: Tei F, Nicola S, Benincasa P, editors. Advances in research on fertilization management of vegetable crops. Advances in Olericulture. Cham.: Springer; 2017. DOI: 10.1007/978-3-319-53626-2_7

[56] Létourneau G, Caron J, Anderson L, Cormier J. Matric potential-based irrigation management of field-grown strawberry: Effects on yield and water use efficiency. Agricultural Water Management. 2015;**161**:102-113

[57] Abul-Soud M, Mohammed MH, Emam MSA, Maharik ZY. Sustainable ecology strawberry production via different soilless culture techniques under urban conditions. Egyptian Journal of Horticulture. 2020;**47**(2):205-218

[58] Gallace N, Lieten P. Potassium-calcium ratio of the nutrient solution: Implications for fruit quality in June-bearing and day-neutral strawberry cultivars. Acta Horticulturae. 2018;**1217**:405-410

[59] Othman Y, Bataineh K, Al-Ajlouni M, Alsmairat N, Ayad J, Shiyab S, et al. Soilless culture: Management of growing substrate, water, nutrient, salinity, microorganism and product quality. Fresenius Environmental Bulletin. 2019;**28**(4):3249-3260

[60] Nestby R, Lieten F, Pivot D, Raynal-Lacroix C, Tagliavini M. Influence of mineral nutrients on strawberry fruit quality and their accumulation in plant organs: A review. International Journal of Fruit Science. 2004;**5**(1):141-158

[61] Jarosz Z, Konopinska J. Effect of substrate type and nitrogen fertilization upon yielding and chemical composition of "elsanta" strawberry cultivar grown in unheated foil tunnel. Acta Scientiarum Polonorum, Hortorum Cultus. 2010;**9**(1):87-96

[62] Tang X, Li Y, Fang M, Li W, Hong Y, Li Y. Effects of different water storage and fertilizer retention substrates on growth, yield and quality of strawberry. Agronomy. 2024;**14**:205. DOI: 10.3390/agronomy14010205

[63] Roosta HR, Sajjadinia A, Rahimi A, Schjoerring JK. Responses of cucumber plant to NH_4^+ and NO_3^- nutrition: The relative addition rate technique vs cultivation at constant nitrogen concentration. Scientia Horticulturae. 2009;**121**(4):397-403

[64] Samec D, Maretić M, Lugarić I, Mešić A, Salopek-sondi B, Duralija B. Assessment of the differences in the physical, chemical and phytochemical properties of four strawberry cultivars using principal component analysis. Food Chemistry. 2016;**194**(1):828-834

[65] Sarooshi RA, Cresswell GC. Effects of hydroponic solution composition, electrical conductivity and plant spacing on yield and quality of strawberries. Australian Journal of Experimental Agriculture. 1994;**34**:529-535

[66] Li YL, Stanghellini C, Challa H. Effect of electrical conductivity and transpiration on production of greenhouse tomato (*Lycopersicon esculentum* L.). Scientia Horticulturae. 2001;**88**:11-29

[67] Nestby R, Retamales JB. Diagnosis and management of nutritional constraints in berries. In: Srivastava AK, Hu C, editors. Fruit Crops: Diagnosis and Management of Nutrient Constraints. Amsterdam, Netherlands: Elsevier; 2020. pp. 567-582

[68] Hancock J, Retamales J. Blueberry taxonomy and breeding. In: Blueberries. Wallingford: CABI; 2012. pp. 19-49. DOI: 10.1079/9781845939045.0019

[69] Nunez GH, Zapien M, Phillips DA. Introduction to Southern Highbush Blueberry Cultivation in Containers. Vol. 2024, No. 1. Florida, USA: EDIS, UF/IFAS; 2024. DOI: 10.32473/edis-HS1476-2024

[70] Doyle JW, Nambeesan SU, Malladi A. Physiology of nitrogen and calcium nutrition in blueberry (*Vaccinium sp.*). Agronomy. 2021;**11**:765. DOI: 10.3390/agronomy11040765

[71] Wang X, Wang Y, Wang J. Effect of different solid medium on blueberry soilless culture. Advances in Engineering Research. 2017;**120**:2099-2102

[72] Yasin M, Jabran K, Afzal I, Iqbal S, Nawaz MA, Mahmood A, et al. Industrial sawdust waste: An alternative to soilless substrate for garlic (*Allium sativum* L.). Journal of Applied Research on Medicinal and Aromatic Plants. 2020;**18**:100252

[73] Hortidaily. Developments in Substrates for the Soft Fruit Sector. 2015. Available from: http://www.hortidaily.com/article/19284/Developments-in-substrates-for-the-soft-fruit-sector

[74] Clark MJ, Zheng Y. Fertilization methods for organic and conventional potted blueberry plants. HortScience. 2020;**55**:304-309

[75] Evans MR, Konduru S, Stamps RH. Source variation in physical and chemical properties of coconut coir dust. HortScience. 1996;**31**:965-967

[76] Lemaire F. Physical, chemical and biological properties of growing medium. Acta Horticulturae. 1995;**396**:273-284. DOI: 10.17660/actahortic.1995.396.33

[77] De Silva A, Patterson K, Rothrock C, McNew R. Phytophthora root rot of blueberry increases with frequency of flooding. HortScience. 1999;**34**(693):695

[78] Machado RMA, Bryla DR, Vargas O. Effects of salinity induced by ammonium sulfate fertilizer on root and shoot growth in highbush blueberry. Acta Horticulturae. 2014;**1017**:407-414

[79] Frías-Ortega CE, Alejo-Santiago G, Bugarín-Montoya R, Aburto-González CA, Juárez-Rosete CR, Urbina-Sánchez E, et al. Nutritive solution concentration and its relationship with blueberry production and quality. Ciencia y Tecnología Agropecuaria. 2020;**21**(3):e1296

[80] Li T, Bi G. Container production of southern highbush blueberries using high tunnels. HortScience. 2019;**54**:267-274

[81] Fulcher A, Gauthier NW, Klingeman WE, Hale F, White SA. Blueberry culture and pest, disease, and abiotic disorder management during nursery production in the southeastern US: A review. Journal of Environmental Horticulture. 2015;**33**(33):47

[82] Voogt W, Van DP, Douven F, Van Der Maas R. Development of a soilless growing system for blueberries (*Vaccinium corymbosum*): Nutrient demand and nutrient solution. Acta Horticulturae. 2014;**1017**:215-221

[83] Fang Y, Williamson J, Darnell R, Li Y, Liu G. Optimizing nitrogen fertigation rates for young southern highbush blueberry. Agronomy. 2020;**10**:1-14

[84] Whidden A. Commercial blueberry production methods in Hillsborough County. Proceedings of the Florida State Horticultural Society. 2008;**121**:6-37

[85] Cantliffe DJ. Pre- and postharvest practices for improved vegetable transplant quality. HortTechnology. 1993;**3**:415-416

[86] Wilber WL, Williamson JG. Effects of fertilizer rate on growth and fruiting of containerized southern highbush blueberry. HortScience. 2008;**43**:143-145

[87] Fang Y, Nunez G, Fisher P, Munoz PR. Effect of container size, substrate composition, and genotype on growth and fruit quality of young southern highbush blueberry in a container-based intensive production system. Scientia Horticulturae. 2022;**302**:111149

[88] Yaeger T, Fare D, Lea-Cox J, Ruter J, Bilderback T, Gilliam C. Best Management Practices: Guide for Producing Container-Grown Plants. Marietta, GA, USA: Southern Nursery Association; 2007. pp. 1-7

[89] Owen JS, Altland JE. Container height and Douglas fir bark texture

affect substrate physical properties. HortScience. 2008;**43**:505-508

[90] Kang S, van Iersel MW, Kim J. Plant root growth affects FDR soil moisture sensor calibration. Scientia Horticulturae. 2019;**252**:208-211

[91] Ciordia M, Díaz MB, García JC. Blueberry culture both in pots and under Italian-type tunnels. Acta Horticulturae. 2002;**574**:123-127

[92] Salazar-Gutierrez M, Lawrence K. Evaluation of Blueberry Production Using Containers. USA: SRSFC Research Report. Auburn University; 2022

[93] Raviv M, Lieth JH. Soilless Culture Theory and Practice. San Diego, CA: Elsevier; 2008

[94] Mathers HM, Yeager TH, Case LT. Improving irrigation water use in container nurseries. HortTechnology. 2005;**15**:8-12

[95] Million J, Yeager T, Albano J. Consequences of excessive overhead irrigation on runoff during container production of sweet viburnum. Journal of Environmental Horticulture. 2007;**25**:117-125

[96] Fulcher A, LeBude AV, Owen JS, White SA Jr, Beeson RC. The next ten years: Strategic vision of water resources for nursery and greenhouse producers. HortTechnology. 2016;**26**:121-122

[97] Warsaw AL, Fernandez RT, Cregg BM, Andersen JA. Container-grown ornamental plant growth and water runoff nutrient and volume under four irrigation treatments. HortScience. 2009;**44**:1573-1580

[98] Million JB, Yeager TH, Albano JP. Evapotranspiration based irrigation scheduling for container-grown *Viburnum odoratissimum* (L.) Ker Gawl. HortScience. 2010;**45**:1741-1746

[99] Beeson R, Yeager TH. Plant canopy affects sprinkler irrigation application efficiency of container-grown ornamentals. HortScience. 2003;**38**:1373-1377

[100] Chappell M, Dove SK, van Iersel MW, Thomas PA, Ruter J. Implementation of wireless sensor networks for irrigation control in three container nurseries. HortTechnology. 2013;**23**:747-753

[101] Smith E, Porter W, Hawkins G, Harris G Jr. Blueberry Irrigation Water Quality. Athens, GA, USA: University of Georgia Cooperative Extension; 2016

[102] Bryla DR, Strik BC. Nutrient requirements, leaf tissue standards, and new options for fertigation of northern highbush blueberry. HortTechnology. 2015;**25**(4):464-470

[103] Schreiber MJ, Nunez GH. Calcium carbonate can be used to manage soilless substrate pH for blueberry production. Horticulturae. 2021;**7**:74

[104] Tamir G, Afik G, Zilkah S, Dai N, Bar-Tal A. The use of increasing proportions of $N\text{-}NH_4^+$ among the total applied inorganic N to improve acidification and the nutritional status and performance of blueberry plants in soilless culture. Scientia Horticulturae (Amsterdam). 2020;**276**:109754. DOI: 10.1016/j.scienta.2020.109754

[105] Balboa K, Ballesteros GI, Molina-Montenegro MA. Integration of physiological and molecular traits would help to improve the insights of drought resistance in highbush blueberry cultivars. Plants. 2020;**9**:1-16

[106] Lobos GA, Hancock JF. Breeding blueberries for a changing global environment: A review. Frontiers in Plant Science. 2015;**6**:1-14

[107] Álvarez Y, Oliva M, Collazos R, Vilca N, Huaman E. Desempeño agronómico de cuatro variedades de arándano (*Vaccinium corymbosum* L.) cultivadas en diferentes sustratos y pisos altitudinales. Bioagro. 2020;**32**:187-194

[108] Valentinuzzi F, Pii Y, Mimmo T, Savini G, Curzel S, Cesco S. Fertilization strategies as a tool to modify the organoleptic properties of raspberry (*Rubus idaeus* L.) fruits. Scientia Horticulturae. 2018;**240**:205-212

[109] Frías-Moreno MN, Parra-Quezada RA, González-Aguilar G, Ruíz-Canizales J, Molina-Corral FJ, Sepulveda DR, et al. Quality, bioactive compounds, antioxidant capacity, and enzymes of raspberries at different maturity stages, effects of organic vs. conventional fertilization. Foods. 2021;**10**:953

[110] Hanson E. Substrate Culture of Raspberries., Department of Horticulture. USA: Michigan State University. Annual Report to NARBA; 2017

[111] Hart J, Strik BC, Rempel H. Caneberries; Nutrient Management Guide. Corvallis, OR, USA: Oregon State University, Extension Service; 2006; EM 8903-E

[112] Sonneveld C. Mineralenbalansen bij kasteelten. In: NNI Wageningen. Vol. 432. Meststoffen; 1993. pp. 44-49

[113] Bryla DR, Shireman AD, Machado RMA. Effects of method and level of nitrogen fertilizer application on soil pH, electrical conductivity, and availability of ammonium and nitrate in blueberry. Acta Horticulturae. 2010;**868**(95):101

[114] Gerbrandt EM, Mouritzen C, Sweeney M. Foliar calcium corrects a deficiency causing green fruit drop in 'Draper' highbush blueberry (*Vaccinium corymbosum* L.). Agriculture. 2019;**9**(3):63

[115] Balawejder M, Matłok N, Piechowiak T, Szostek M, Kapusta I, Niemiec M, et al. The modification of substrate in the soilless cultivation of raspberries (*Rubus Idaeus L.*) as a factor stimulating the biosynthesis of selected bioactive compounds in fruits. Molecules. 2023;**28**:118

[116] Carlen C, Ançay A, Christ B. Optimization of the root environment for raspberry production on substrate. Acta Horticulturae. 2020;**1277**:283-286

[117] Koester K, Pritts M. Greenhouse Raspberry Production Guide. Project Report. USA: Cornell University; 2003. Available from: www.fruit.cornell.edu/berry.html

[118] Talluto G, Farina V, Lo BR. Growth, fruit yield and quality of 'Golden delicious' apple trees under fixed partial root zone drying. Journal of Applied Horticultre. 2007;**9**(01):50-55

[119] Lepaja K, Kullaj E, Lepaja L, Zajmi A. Effect of PRD on raspberry cultivation in pots. Acta Horticulturae. 2021;**1308**:93-98. DOI: 10.17660/ActaHortic.2021.1308.15

[120] Dry PR, Loveys BR. Factors influencing grapevine vigour and the potential for control with partial root zone drying. Australian Journal of Grape and Wine Research. 1998;**4**(3):140-148

[121] Dodd IC. Root-to-shoot signaling: Assessing the roles of «up» in the

up and down world of long-distance signaling in plants. Plant and Soil. 2005;**274**(1-2):251-270

[122] Grant OM, Stoll M, Jones HG. Partial rootzone drying does not affect fruit yield of raspberries. The Journal of Horticultural Science and Biotechnology. 2004;**79**(1):125-130

[123] Bolda M, Gaskell M, Mitcham E, Cahn M. Fresh Market Caneberry Production Manual. Richmond, California: University of California Agriculture and Natural resources; 2012. Publication 3525

[124] Pitsioudis A, Latet G, Meesters P. Out of season production of raspberries. Acta Horticulturae. 2002;**585**:555-560

[125] Strik BC, Bryla DR. Uptake and partitioning of nutrients in blackberry and raspberry and evaluating plant nutrient status for accurate assessment of fertilizer requirements. HortTechnology. 2015;**25**:452-459

[126] Koron D, Munda A, Modic Š, Poje T, Leskovšek R, Vončina A, et al. Smernice integriranega varstva maline. Kmetijski inštitut Slovenije. 2017;**24**:7-9

[127] Bushway L, Pritts M, Handley D. Raspberry and Blackberry Production Guide for Northeast, Midwest and Eastern Canada. Ithaca. New York: Plant and Life Sciences Publishing (PALS) Cooperative Extension; 2008. Available from: https://ecommons.cornell.edu/handle/1813/66930

Chapter 7

Perspective Chapter: Hydroponic Farming for Food and Nutrition Security – Challenges and Prospects

Lemma Tessema and Zebenay Dagne

Abstract

The continuous growth of the global population, coupled with water shortages, climate change, pest and disease pressures, and urbanization, poses significant risks to the agriculture sector's ability to feed the world's 8 billion people. The limited natural resources, alongside increasing human needs and the ambitions of the United Nations' Sustainable Development Goals (SDGs), necessitate modernized agriculture that can produce high quantities of quality food. Hydroponics—a vertical, soilless crop production technique for food, animal feed, and ornamental purposes—emerges as a valuable alternative in modern agriculture, especially in areas where critical production elements like land and water are scarce. This innovative approach offers numerous advantages over conventional farming, including efficient water and space utilization, year-round production, reduced environmental impact, and enhanced environmental control through ongoing technological advancements. Due to its space-efficient nature, many farmers are increasingly adopting hydroponics for vegetable and fodder production. Therefore, hydroponic technology provides a sustainable solution for food and feed production, optimizes space use, conserves water, controls diseases, and addresses the multifaceted challenges of conventional agriculture, contributing to the achievement of the United Nations' SDGs.

Keywords: hydroponic technology, soilless cultivation, SDGs, environment friendly, conventional farming

1. Introduction

The continuous growth of population, shortage of water supply, climate change, pest and disease pressure, and urbanization put the agriculture sector at risk to feed the world's 8 billion lives. The scarcity of natural resources, combined with the increasing demands of humanity, poses a major challenge to the United Nations' Sustainable Development Goals (SDGs), which are currently off track in meeting the targets of the 2030 Agenda [1, 2]. Crop production in Africa including Ethiopia is characterized by low productivity due to poor soil fertility, low input usage, climate change, pests and diseases as well as various environmental constraints [3]. In order to devise a strategy to boost crop productivity where food insecurity is most prevalent and chronic, adopting sustainable intensification through agricultural technologies and innovations is very critical. Modernizing the agriculture sector in this regard would play a key role that enables to intensively produce high quantity and -quality produce and to benefit the community from the sector. In addition to considering how successfully new agricultural technologies contribute to the sustainability and resilience of agriculture, their effectiveness in meeting the demands and needs of family farmers for increased revenue and productivity should also be taken into consideration [4].

Globally, three significant challenges—food security, nutrition security, and climate security—confront the twenty-first century and are closely tied to traditional farming and food production systems, which hinder efforts to achieve the United Nations' Sustainable Development Goals. The rapid increase in the global population, combined with decreasing arable land, soil erosion, and limited freshwater resources, disrupts environmental stability. Addressing these interrelated issues is essential and can be partially mitigated through climate-smart agricultural practices, such as using renewable energy in farming, diversifying crops, rotating crops, and embracing organic methods. However, there is an urgent need to transition to climate-resilient farming techniques. Soilless indoor cultivation technology offers a promising long-term sustainable strategy. By adopting soilless methods, farmers can increase agricultural productivity with less land, optimize water use efficiency, reduce reliance on harmful chemicals, and enhance ecological sustainability [5, 6].

Hydroponics, a vertical, soilless cultivation technique for food, animal feed, and ornamental purposes, is an alternative method expected to reduce the sector's burden and gain an importance by the modern agriculture, especially where the most critical production elements, land and water, are scarce [7]. This agricultural innovation has numerous advantages over the conventional farming system, including efficient water and space utilization, year-round production and environment friendly production, higher crop yield, and profitability [2, 8, 9]. Furthermore, hydroponic farming is considered a clean and environment friendly vegetable production technique due to its minimum risk of soil-borne disease and pest infection to the crops, thereby reducing chemical spray and less toxicity to the environment [10].

Hydroponic technology offers a sustainable solution for food and feed production by enabling accelerated year-round cultivation to address the multifaceted challenges of conventional farming [9]. Despite its numerous benefits, the higher initial investment, total operating costs, and increased energy consumption compared to traditional farming hinder its adoption and scalability, particularly among smallholder farmers in developing countries, where technological advancements and financial resources are often limited [8, 9]. However, hydroponic systems can be constructed using simple, locally available materials, and they can operate without external energy sources unless automated, which typically requires greater investment and energy consumption [11]. This chapter aims to provide insights into the challenges and prospects of hydroponic farming for food and nutrition security, as well as its environmental and economic roles within the agri-food system amid climate change and the growing global population.

2. Hydroponic farming

Hydroponic farming is among soilless cultivation techniques that provide an opportunity to the agriculture sector, especially in areas where there are serious soil degradation and/or land and water scarcity problems [12]. This technology also provides an opportunity to feed the ever-growing populations in cities by enabling year-round production of food and feed crops with limited input and water wastage [13]. Soilless farming, including hydroponics, reduces the threat of yield loss due to soil-borne pests and disease, soil salinity problems, and other interconnected yield impeding factors. Hydroponic technology also minimizes solid wastes and less energy consumption for food and feed production, when the drip system is only working with gradient without any power to pump the nutrient solution to plant roots [14].

2.1 Principles and basic concepts of hydroponic farming

With hydroponics, a nutrient-rich solution is provided into plant roots to facilitate soilless plant growth. When compared to conventional soil-based approaches, this technique for growing plants has a number of advantages. For instance, hydroponic systems can offer larger crops in less space and require less water than soil-based systems. Additionally, the ever-increasing climate change induced environmental issues, there is a need to find ways of resilient crop production technologies to boost crop productivity. In this regard, hydroponic farming has gained interest, as it reduces dependency on agricultural land and pesticide for crop pest control, as well as its merit to produce food and feed with poor soil and scarce water, especially to feed highly populated urban dwellers [15].

There are different types of hydroponics mainly based on their substrate type used, technological advancement, functionality, usage, and types/purpose of crop grown under hydroponic system (**Table 1**) [16]. As far as plant-growing substrate medium is concerned, there are different hydroponic systems. For example, there is a type of hydroponic called "sand hydroponics" at Holetta Agricultural Research Center established in 2013 for seed tuber production. The sand medium is disinfected with sodium hypochlorite and free from any pathogens, it is just used as a growing medium for potato early generation seed production (**Figure 1**).

2.2 Media and nutrient handling in hydroponic cultivation

In hydroponic farming, proper nutrient management is crucial, as crops do not access macro- and micronutrients from the soil, unlike in conventional farming, where many essential nutrients are readily available. Each crop has specific nutritional requirements, and the type and quantity of nutrients used in hydroponics must be carefully managed [17, 18]. Typically, 17 essential plant nutrients are supplied in the form of inorganic fertilizers for hydroponic systems, though variations may occur based on local availability and accessibility of fertilizers [19]. Technicians working in hydroponic greenhouses should regularly monitor nutrient levels by measuring concentrations in the solution to identify any imbalances or deficiencies. Adjusting nutrient levels according to the crop's requirements is essential to ensure that plants receive adequate nutrient solutions. In this context, managing nutrient pH and electrical conductivity (EC) is critical. Proper adjustments to pH and EC are necessary to meet the specific needs of the crops being produced [18].

In some cases, the strength of the nutrient solution and the stage of plant growth influence variations in pH, electrical conductivity (EC), and mineral concentrations more than the specific type of nutrient solution used [20]. The nutritional requirements of crops grown in hydroponics may differ in terms of nutrient concentration and solution pH. For example, research on hydroponic potato mint production suggests that a nutrient solution with a strength of 0.5 to 1.0 milliequivalent per liter supports optimal crop growth and tuber yield [20]. At the Holetta Agricultural Research Center, we utilize nutrient solutions based on elemental requirements for hydroponic seed tuber production, adapted from

Hydroponic farming systems	Description and principles	Reference
Wick system	The simplest type of hydroponic system where a wick draws water and nutrients from a reservoir to the plant roots. The plant is cultivated in a container with a growing medium such as perlite or coconut coir.	[15]
Deep water culture (DWC)	The plant roots are submerged in a nutrient solution that is aerated by an air pump. The plant roots are not in contact with any growing medium and air pump provides oxygen to the roots and helps to circulate the nutrient solution.	[16]
Nutrient film technique (NFT)	A thin film of nutrient solution flows through a channel or tube where the plant roots are exposed to contact with the liquid. A thin film of nutrient solution flows through a channel or tube where the plant's bare roots are allowed to grow and come in contact with it. The nutrient solution comes from the reservoir to the top of the channel and flows down to the bottom by gravity. The roots absorb the necessary nutrients, and excess water drains back into the reservoir.	[15]
Ebb and flow (or flood and drain)	The roots of the plants growing on a trail are periodically submerged in the nutrient-rich solution, which is left to empty back into the reservoir for future use. In ebb and flow (or flood and drain), the plant roots are periodically flooded with nutrient solution to supply the plants with water and nutrients. The solution is slowly drained back into the reservoir, and this cycle is repeated several times a day.	[15]
Drip system	A pump delivers a nutrient solution to each plant through a drip emitter. In the drip system, the nutrient solution is dripped onto the base of each plant through a small tube or emitter. The excess solution drains back into the reservoir or drains it out form root zone.	[14, 15]
Aeroponics	The plant roots are sprayed with a fine mist of nutrient-rich solution to plant roots. The nutrient solution is sprayed onto the roots in a fine mist using high-pressure misters. The excess nutrient solution is collected and recycled back into the reservoir.	[16–18]
Aero hydroponics	Design to improve the root zone environment of aeroponics by maintain root contact with nutrient solution in the lower part of the beds.	[16]
Sand hydroponics	Works with drippers without any energy source, only with slope variation from nutrient source to planting boxes. The name is taken from the substrate medium 'sand' and used for quality seed tuber production.	[14]

Table 1.
Various hydroponic systems and their applications for crop production.

Otazu [17], formulations developed for aeroponics [18]. The nutrient solutions for both aeroponic and hydroponic farming at Holeta are adjusted to an EC of 1.5 mS/cm and a pH of 6.8 [18], although a pH range of 5.0–5.6 is generally recommended for hydroponic crop production. In addition to the 17 essential mineral elements necessary for plant growth, the inclusion of elements, such as cerium, cobalt, iodine, aluminum, silicon, titanium, lanthanum, selenium, sodium, and vanadium, is increasingly recognized as important bio-stimulants that enhance plant growth and improve stress tolerance [21].

In hydroponic farming, different plant growth-supporting media could be used based on their relevance and availability. As substrates are key elements in hydroponic farming, the type and characteristics of the substrate used for hydroponics may vary based on their availability and accessibility. More likely, substrates, which allow the growers to easily control the root environment, are preferable. Plant-growing substrates

Figure 1.
Potato mini-tuber production through sand hydroponics at Holetta Agricultural Research Center, Ethiopia (©Holetta potato research program).

in hydroponics play a pivotal role by supporting the plants and facilitating ideal nutrient movement as well as enhanced aeration to the plant root zone. The plant-growing substrates in hydroponics are used for these major advantages; (i) act as a reservoir, holding moisture and nutrients around the roots of plants, ensuring nutrients remain dissolved and available for absorption; (ii) form pore spaces around roots to facilitate oxygenation, enabling roots to respire effectively; and (iii) secure roots in place and offer structural support, helping plants grow upright. To mention, some common growth substrates in hydroponics system are: peat moss, coconut coir, perlite, vermiculite, or others. If these plant-supporting substrates are not available or are too costly, well-disinfected coarse sand can also be used for hydroponic crop production [14].

2.3 Crop management

In hydroponic farming, crop management varies based on the type and nature of the cultivating crop. However, the technicians working in hydroponic farming need to be skillful in crop physiology and agronomy as well as crop-nutrient requirement. As far as crop type is concerned, hydroponic technology can be used to cultivate a broad variety of crops, such as flowers, vegetables, fruits, forages, cereals, medicinal and ornamental plants, though small- to medium-sized crops are more appropriate economically and agronomically (**Table 2**) [13, 18, 24]. Crop handling under protected production system needs serious follow-up and instant action. If the technician under greenhouse observes any diseased or wilted plants, the infected plant should immediately be removed since the infected plant will reinfect healthy plants through nutrient recycling under individual plant roots. In hydroponic production system, water-borne diseases pose a significant risk to our crops and controlling water temperature is also very crucial since roots are oxygen deprived and pathogens can flourish under

Type of crop	Purpose of hydroponic used	Type of hydroponic used	Reference
Potato	The production of high quality potato mini-tubers	Three hydroponic systems (aeroponics, aero hydroponics, and deep-water culture)	[16]
Cereals	Forage for animal nutrition	Hydroponic and aquaponic techniques	[22]
Vegetables	Lettuce for food and nutrition		
Potato	Potato production	Nutrient film technique (NFT)	[23]
Lettuce	Home consumption and income source	Different types of vertical farming	[13]
Spinach	Home consumption and income source	Different types of vertical farming	[13]
Bell pepper	Home consumption and income source	Different types of vertical farming	[13]
Tomatoes	Home consumption and income source	Different types of vertical farming	[13]
Maize	Green fodder for animals	Nutrient film technique (NFT)	[24]

Table 2.
Cultivation of food and feed crops under different hydroponic systems.

nutrient solution [22]. Implementing restricted sanitation rule at each hydroponic system is critical to reduce disease and pest spread. Vertical farming system requires intensive crop management and follow-up. With the use of hydroponic technology, growers can manage crop development in protected environments using machine learning techniques to increase production and yield [25].

2.4 Environmental impacts and economic viability of hydroponic farming

Aside easy management of crops under hydroponic farming, it ensures zero growth of weeds and no herbicide spray, contributing to less environmental pollution by herbicide residues [19]. The study of Paturu and Varadarajan [26] evaluated different farming models using the life cycle assessment (LCA) method in lettuce crop and suggested greenhouse farming system is more environmentally sustainable than indoor farming as well as traditional farming methods. This method also realizes greenhouse farming performs well in greenhouse gas emission, human toxicity, and fossil fuel usage compared to other crop production models. The study of Souza [27] argues that hydroponics can be an attractive alternative to produce food in small properties. Likewise, hydroponics-based crop production also benefited many million households during the COVID-19 pandemic crisis, when people were out of their jobs and they could generate income through vegetable production at home [28]. So, space-efficient production systems like hydroponic contributed a lot with multiple social and economic advantages to the community, especially for urban dwellers in the Global South. In the context of urban agriculture, hydroponics could be effectively used in cities with simpler and less expensive technology, particularly in areas of extreme poverty, as a means of dispersing vegetables for personal consumption, supporting the income of the family or community, and establishing microbusinesses that will create jobs to youth and women [28].

However, Kalantari et al. [29] have appealed that vertical farming is not a feasible alternative for cultivating extensive or low-value crops. From a sustainable and environment friendly crop production perspective, hydroponic technology can be utilized to treat contaminated water. This approach helps to reduce the pollutant load on the environment, as the contaminants can serve as essential nutrients for the plants [30]. Hydroponic farming minimizes environmental deterioration by producing less runoff and soil erosion than traditional farming methods since it provides high water saving opportunity to growers [31]. Furthermore, hydroponic farming reduces carbon emissions and soil compaction by doing away with the need for heavyweight machinery. This system recirculates water, reducing overall water consumption and minimizing wastage [2, 9]. Nowadays, water scarcity, nutrient-depleted soils, and pollution remain significant global challenges, which are expected to exacerbate with the growing global population, especially in urban areas [32]. Hydroponic system offers an opportunity in reusing municipal wastewater for crop production in many urban areas of the world. This contributes significantly to enhancing food and nutrition sovereignty and ecological preservation. Hydroponic systems are also a good choice for treating wastewater since they can lessen the health concerns that come with direct contact with untreated wastewater for farmers, harvested crops, and consumers [32]. The study on lettuce production in decoupled aquaponic system in fish water saved 62.8% mineral fertilizer and totally substituted the required water for nutrient solution. This is because decoupled aquaponics uses less energy to produce fertilizer and reduced fertilizer use and helped a lettuce-producing site's yearly greenhouse gas emissions per hectare by 72% [33].

2.5 Nutritional aspects of hydroponically grown crops

Nutritionally, hydroponically produced foods and feeds are palatable and fresh to consume. Hydroponic farming provides an advantage of producing and delivering fresh vegetables and green forages to the local community particularly in urban dwellers, where people are consuming foods that traveled from remote areas of the country for days or weeks under transportation. In most cases, fresh produces of vegetables and fruits harvested from hydroponic farming are nutritionally safe to the consumers and the fresh forage harvests are palatable to animals. Hydroponically grown fodder is nutritious and free from contamination, which can improve animal health and reproductive efficiency to be ready to feed our animals within 7 days [22]. Hydroponic lettuce had increased moisture leading to lower bioactive compounds and had lower antioxidant capacity than soil-grown lettuce [34]. According to the health risk index (HRI < 1) and targeted hazardous quotient (THQ < 1) investigated by Zhu et al. [31], there was little chance that heavy metals would pose a health concern in hydroponically grown lettuce produce.

Since hydroponic farming allows growers to harvest fresh produce, these fresh produces have abundant bioactive compounds, including vitamins, minerals, and phytochemicals with good nutritional qualities [35]. From the human health perspectives, these fresh microgreens can neutralize free radicals and help to prevent damage caused by oxidative stress due to the presence of vitamin C, phytochemicals like carotenoids and phenolics, certain minerals including copper, zinc, and selenium [35]. The study at the University of Nevada, Reno, researchers compared hydroponically grown strawberries and raspberries with those grown in soil. They found that hydroponically grown strawberries had notably higher amounts of vitamin C, vitamin E, and total polyphenolic compounds, but lower levels of fructose and glucose compared to soil-grown strawberries and raspberries. Compared to soil-grown strawberries and raspberries, hydroponically grown berries produced a higher yield, comparable or superior nutritional quality and taste preferences [36].

3. Challenges related to hydroponic farming

Hydroponic farming with its multiple benefits, there are limitations in its implementation for improved crop production and productivity, such as high investment costs and lack of adequate knowledge, to mention some [13]. The study of Wagner et al. [2] also noted that initial costs of hydroponics, such as greenhouse construction, costs of fertilizers, and other hydroponic equipment costs, are mentioned as a challenge for adoption of hydroponic farming technology in many developing countries. Al-Kodmany [37] similarly concluded that economic visibility and lack of expertise remain the major obstacles in the path to implementing the vertical farm. Pests and diseases are other challenges in hydroponic farming under screenhouse. One of the most common culprits behind root rot in hydroponics is a group of water-borne fungi, particularly *Pythium* [22]. These microscopic pathogens thrive in moist, oxygen-poor environments, making hydroponic systems an ideal breeding ground. *Pythium* attacks plant roots, causing them to turn soft, brown, and mushy. As the infection spreads, the roots lose their ability to absorb nutrients, leading to the decline of the entire plant [22]. The most common causes of root rot in hydroponic farming are due to: insufficient oxygen, high water temperatures, and contaminated equipment and nutrient solution. Common challenges in hydroponic farming include, but not limited to (i) nutrient deficiencies or excesses, (ii) pH fluctuations, (iii) pest and disease infestations, and (iv) well-trained manpower who can properly operate the hydroponic system. To cope up with possible pest and disease control under hydroponic system, integrated pest management (IPM) strategies, including biological controls, monitoring, and sanitation, are essential.

A comparative study in Yuma, Arizona, examining the land, water, and energy requirements of hydroponic versus conventional agriculture highlighted significant advantages of hydroponics. Hydroponic systems demonstrated an 11-fold increase in yield while consuming less water; however, they required 82 times more energy than conventional lettuce production [9]. To improve water efficiency and enhance the sustainability of hydroponic farming, additional technologies such as drip irrigation and fogging systems are necessary, particularly in water-scarce regions. Furthermore, the sustainability and accessibility of hydroponic technology can be enhanced by addressing its energy demands through renewable sources, such as solar, geothermal, or wind power. This is especially critical in developing countries, where the cost of electric power can be prohibitively high. By leveraging renewable energy, hydroponics can become a more viable and sustainable option for farmers facing water scarcity.

4. Prospects of the hydroponic farming

The hydroponic farming system has a potential to pave into smart agriculture to improve food and nutrition security in the agriculture sector. Hydroponic farming, with its time and space efficiency, production of quality green fodder for animals was possible within 7 days [38], which is unlikely with conventional farming systems. As a modern agricultural technology, hydroponic farming promises to provide high-quality, healthy, fresh, residue-free vegetables and fruits to the locality [10]. Additionally, research indicates that hydroponic farming could become a significant method for commercial vegetable production, despite its higher initial investment and the need for advanced technology [8].

In the near future, the world's hydroponic farming market will worsen reaching a value of $25.1 billion encompassing various participants, such as food manufactures, crop cultivators, hydroponic system manufacturers, and suppliers of hydroponic components [23]. Future research may involve larger-scale experiments to determine if the hydroponic system can satisfy market demand of the future [39]. As the result, this modern agriculture technology opens doors for researchers to investigate and generate more information on the economic and environmental impact of hydroponics-based urban farming models in the arena of climate-smart agriculture for resilient agri-food systems. Hence, hydroponic farming would provide opportunities for research and development for scientists to perform multidisciplinary research on crop-nutrient requirement, crop physiology, pathology, and engineering until availing adequate information for controlled environment production system suitable for different agro-ecological regions and crop diversity of the world [23].

Hydroponics offers the potential to rebalance industrial development and agriculture, reintroduce farming into urban areas as an alternative livelihood for residents, and help alleviate sustainable water scarcity issues [39]. Hydroponic system was observed to be more sensitive to effects of environmental temperature than traditional cultivation due to higher soil buffering capacity. The technology also enabled to produce higher yield (134%) and higher water productivity (50%) under greenhouse lettuce production experiment [8]. The future of hydroponic farming holds great promise, with the potential to fundamentally transform food production. Hydroponic farming not only delivers high yields in productive soils but also enables ample food production in areas with degraded soils [19]. So, hydroponic farming attracts both technologists and the consumers due to its multiple advantages [23]. As per the study in East Africa by Gumisiriza et al. [13], hydroponics has a potential to increase food security for urban dwellers where populations are highly dense with putting efforts to reduce its initial cost and sensitizing its framing system suitable to majority of urban farmers.

Hydroponic farming has bright future to scale up in the near future due to its unique advantages of efficient resource and space utilization, year-round production, higher crop yield, reduced pest and disease pressure, reduced environmental impact, and precise control over crop growing conditions under screenhouse [2, 8, 14]. Integration of hydroponic farming into urban food production offers the potential to tackle both food security and environmental sustainability, leading to a more resilient food system [40]. In the future, regions experiencing drought may employ desalted salt water in hydroponic systems, enabling food production in coastal areas, deserts, and developing nations where food insecurity and recurrent climate change-induced uncertainties are more prevalent. Meanwhile, astronauts aboard the International Space Station already relish lettuce cultivated hydroponically and hydroponic farming in this regard created an opportunity to conduct future research and innovation in our planet.

5. Conclusion

In recent years, the growing global population and rapid urbanization have significantly increased the demand for sustainable and efficient food production systems, all while facing the complex challenges posed by climate change. Traditional agriculture encounters numerous obstacles, including limited arable land, water scarcity, climate change, and soil degradation, which threaten food and nutrition security. Consequently, there is an urgent need to explore innovative agricultural practices that can overcome these limitations and ensure a stable food supply in line with population growth dynamics. Hydroponic farming emerges as a vital solution to global challenges, such as food security, urbanization, and environmental preservation, thanks to its advantages in sustainability, efficiency, and flexibility. With ongoing technological advancements and increasing support from both public and private sectors, hydroponic agriculture is poised to play an increasingly integral role in modern agricultural practices. This method enables crop cultivation in harsh climates and limited spaces, making it essential for ensuring food security in an era where climate change threatens sustainable crop production. Therefore, it is imperative to scale up and integrate hydroponic technology into current agricultural systems, updating these systems with the latest technological advancements to promote environmentally friendly food production and enhance sustainable agri-food systems.

Author details

Lemma Tessema* and Zebenay Dagne
Holetta Agricultural Research Center, Ethiopian Institute of Agricultural Research, Addis Ababa, Ethiopia

*Address all correspondence to: lematessema@gmail.com

References

[1] UN (United Nations General Assembly Economic and Social Council). Progress towards the Sustainable Development Goals. Report of the Secretary-General2024. p. 26. Available from: https://unstats.un.org/sdgs/files/report/2024/SG-SDG-Progress-Report-2024 [Accessed: 10 June 2024]

[2] Wagner N, Morrish D, Juarez M. Identifying the influential factors, benefits and challenges of hydroponic shipping container farm businesses: A snapshot of farmers' perceptions. Renewable Agriculture and Food Systems. 2021;**36**(6):519-526. DOI: 10.1017/S1742170521000211

[3] Tadele Z. Raising crop productivity in Africa through intensification. Agronomy. 2017;**7**:22. DOI: 10.3390/agronomy7010022

[4] UN (United Nations Department of Economic and Social Affairs). Population, Food Security, Nutrition and Sustainable Development. Policy Brief No. 102. 2021. Available from: https://www.un.org/development/desa/dpad/wp-content/uploads/sites/45/publication/PB_102 [Accessed: 20 April 2024]

[5] Banerjee A, Paul K, Varshney A, Nandru R, Badhwar R, Sapre A, et al. Soilless indoor smart agriculture as an emerging enabler technology for food and nutrition security amidst climate change. In: Kumar V, Srivastava AK, Suprasanna P, editors. Plant Nutrition and Food Security in the Era of Climate Change. London: Academic Press; 2022. pp. 179-225. DOI: 10.1016/B978-0-12-822916-3.00004-4

[6] Gebreegziher WG. Soilless culture technology to transform vegetable farming, reduce land pressure and degradation in drylands. Cogent Food & Agriculture. 2023;**9**(2):1-13. DOI: 10.1080/23311932.2023.2265106

[7] Majdalawi MI, Ghanayem AA, Allissa AA, Schlüter S, Tambiah MA, Salman AZ, et al. Economic efficient use of soilless techniques to maximize benefits for farmers. AIMS Agriculture & Food. 2023;**8**(4):1038-1051. DOI: 10.3934/agrfood.2023056

[8] Wang L, Ning S, Zheng W, Guo J, Li Y, Li Y, et al. Performance analysis of two typical greenhouse lettuce production systems: Commercial hydroponic production and traditional soil cultivation. Frontiers in Plant Science. 2023;**14**:1165856. DOI: 10.3389/fpls.2023.1165856

[9] Barbosa GL, Gadelha FD, Kublik N, Proctor A, Reichelm L, Weissinger E, et al. Comparison of land, water, and energy requirements of lettuce grown using hydroponic vs. conventional agricultural methods. International Journal of Environmental Research and Public Health. 2015;**12**(6):6879-6891. DOI: 10.3390/ijerph120606879

[10] Khan S, Purohit A, Vadsaria N. Hydroponics: Current and future state of the art in farming. Journal of Plant Nutrition. 2020;**44**(10):1515-1538. DOI: 10.1080/01904167.2020.1860217

[11] Domingues DS, Hideaki WT, Carlos APC, Suzana LN. Automated system developed to control pH and concentration of nutrient solution evaluated in hydroponic lettuce production. Computers and Electronics in Agriculture. 2012;**84**:53-61. DOI: 10.1016/j.compag.2012.02.006

[12] Sambo P, Nicoletto C, Giro A, Pai Y, Valentinuzzi F, Mimmo T, et al. Hydroponic solutions for soilless production systems: Issues and opportunities in a smart agriculture perspective. Frontiers in Plant Science. 2019;**10**:923. DOI: 10.3389/fpls.2019.00923

[13] Gumisiriza MS, Kabirizi JM, Mugerwa M, Ndakidemi PA, Mbega ER. Can soilless farming feed urban East Africa? An assessment of the benefits and challenges of hydroponics in Uganda and Tanzania. Environmental Challenges. 2022;**6**:100413. DOI: 10.1016/j.envc.2021.100413

[14] Tessema L, Dagne Z. Aeroponics and sand hydroponics: Alternative technologies for pre-basic seed potato production in Ethiopia. Open Agriculture. 2018;**3**(1):444-450. DOI: 10.1515/opag-2018-0049

[15] Rajendran S, Domalachenpa T, Arora H, Li P, Sharma A, Rajauria G. Hydroponics: Exploring innovative sustainable technologies and applications across crop production, with emphasis on potato mini-tuber cultivation. Heliyon. 2024;**10**(5):1-14. DOI: 10.1016/j.heliyon.2024.e26823

[16] Chang DC, Park CS, Kim SY, Lee YB. Growth and tuberization of hydroponically grown potatoes. Potato Research. 2012;**55**:69-81. DOI: 10.1007/s11540-012-9208-7

[17] Otazu V. Manual on Quality Seed Potato Production Using Aeroponics. Lima (Peru): International Potato Center (CIP); 2010. DOI: 10.4160/9789290603924

[18] Tessema L, Chindi A, Gebremedhin W, Solomon A, Shunka E, Seid E. Determination of nutrient solutions for potato (*Solanum tuberosum* L.) seed production under aeroponics production system. Open Agriculture. 2017;**2**(1):155-159. DOI: 10.1515/opag-2017-0015

[19] Kannan M, Elavarasan G, Balamurugan A, Dhanu Siya B, Freed D. Hydroponic farming—A state of art for the future agriculture. Materials Today: Proceedings. 2022;**68**(6):2163-7853. DOI: 10.1016/j.matpr.2022.08.416

[20] Chang D-C, Kim S-Y, Shin K-Y, Cho Y-R, Lee Y-B. Development of a nutrient solution for potato (*Solanum tuberosum* L.) seed tuber production in a closed hydroponic system. Korean Journal of Horticultural Science and Technology. 2000;**18**(1):334-341

[21] Singhal RK, Fahad S, Kumar P, Choyal P, Javed T, Jinger D, et al. Beneficial elements: New players in improving nutrient use efficiency and abiotic stress tolerance. Plant Growth Regulation. 2023;**100**:237-265. DOI: 10.1007/s10725-022-00843-8

[22] Suárez-Cáceres GP, Pérez-Urrestarazu L, Avilés M, Borrero C, Eguíbar JRL, Fernández Cabanás VM. Susceptibility to water-borne plant diseases of hydroponic vs. aquaponics systems. Aquaculture. 2021;**544**:737093. DOI: 10.1016/j.aquaculture.2021.737093

[23] Wheeler RM, Hinkle CR, Mackowiak CL, Sager JC, Knott WM. Potato growth and yield using nutrient film technique (NFT). American Potato Journal. 1990;**67**:177-187. DOI: 10.1007/BF02987070

[24] Reddy B, Harani M. Hydroponics: A sustainable way of green fodder production. Indian Farming. 2023;**73**(02):02-05

[25] Mamatha V, Kavitha JC. Machine learning based crop growth management

in greenhouse environment using hydroponics farming techniques. Measurement: Sensors. 2023;**25**:100665. DOI: 10.1016/j.measen.2023.100665

[26] Paturu P, Varadarajan S. Assessing environmental sustainability by combining product service systems and life cycle perspective: A case study of hydroponic urban farming models in India. Science of the Total Environment. 2024;**927**:172232. DOI: 10.1016/j.scitotenv.2024.172232

[27] Souza SV, Gimenes RMT, Binotto E. Economic viability for deploying hydroponic system in emerging countries: A differentiated risk adjustment proposal. Land Use Policy. 2019;**83**:357-369. DOI: 10.1016/j.landusepol.2019.02.020

[28] Alfredo RD. Peruvian hydroponics: Low-cost options to produce vegetables for south American cities. In: Droege P, editor. Urban and Regional Agriculture. Vol. 2023. London: Academic Press; 2023. pp. 561-594. DOI: 10.1016/B978-0-12-820286-9.00021-2

[29] Kalantari F, Tahir OM, Joni RA, Fatemi E. Opportunities and challenges in sustainability of vertical farming: A review. Journal of Landscape Ecology. 2017;**11**(1):35-60. DOI: 10.1515/jlecol-2017-0016

[30] Mittal R, Bhukal S. Sustainable approach for agriculture and environmental remediation using hydroponics and their perspectives. In: Kumar N, editor. Hydroponics and Environmental Bioremediation. Cham: Springer Water. Springer; 2024. DOI: 10.1007/978-3-031-53258-0_3

[31] Zhu Z, Yogev U, Keesman KJ, Rachmilevitch S, Gross A. Integrated hydroponics systems with anaerobic supernatant and aquaculture effluent in desert regions: Nutrient recovery and benefit analysis. Science of the Total Environment. 2023;**904**:166867. DOI: 10.1016/j.scitotenv.2023.166867

[32] Magwaza ST, Magwaza LS, Odindo AO, Mditshwa A. Hydroponic technology as decentralised system for domestic wastewater treatment and vegetable production in urban agriculture: A review. Science of the Total Environment. 2020;**698**:134154. DOI: 10.1016/j.scitotenv.2019.134154

[33] Monsees H, Suhl J, Paul M, Kloas W, Dannehl D, Würtz S. Lettuce (Lactuca sativa, variety Salanova) production in decoupled aquaponic systems: Same yield and similar quality as in conventional hydroponic systems but drastically reduced greenhouse gas emissions by saving inorganic fertilizer. PLoS One. 2019;**14**(6):e0218368. DOI: 10.1371/journal.pone.0218368

[34] Lie C, Engeseth NJ. Comparison of growth characteristics, functional qualities, and texture of hydroponically grown and soil-grown lettuce. LWT. 2021;**150**:111931. DOI: 10.1016/j.lwt.2021.111931

[35] Zhang Y, Xiao Z, Ager E, Kong L, Tan L. Nutritional quality and health benefits of microgreens, a crop of modern agriculture. Journal of Future Foods. 2021;**1**(1):58-66. DOI: 10.1016/j.jfutfo.2021.07.001

[36] Treftz C, Omaye ST. Comparison between hydroponic and soil systems for growing strawberries in a greenhouse. International Journal of Agricultural Extension. 2015;**03**(03):195-200

[37] Al-Kodmany K. The vertical farm: Exploring applications for peri-urban areas. In: Patnaik S, Sen S, Mahmoud M, editors. Smart Village Technology. Modeling and Optimization

in Science and Technologies. Vol. 17. Cham: Springer; 2020. DOI: 10.1007/978-3-030-37794-6_11

[38] Hassen A, Dawid I. Contribution of hydroponic feed for livestock production and productivity: A review. International Journal of Ground Sediment and Water. 2022;**15**:899-916. DOI: 10.5281/zenodo.5767438

[39] Nazir M, Roy K, Saha A, Saha D. A sustainable holistic approach of hydroponic farming for reclaiming, and rehabilitating wastewater: A review. Water Air Soil Pollution. 2024;**235**:445. DOI: 10.1007/s11270-024-07225-y

[40] Sousa RD, Bragança L, da Silva MV, Oliveira RS. Challenges and solutions for sustainable food systems: The potential of home hydroponics. Sustainability. 2024;**16**:817. DOI: 10.3390/su16020817

Chapter 8

Perspective Chapter: Bioconversion and Bioeconomics of Wastewater from Red Tilapia Aquaculture as a Source of Nutrient (Nitrate) on the Green Mustard Growth with Recirculation and Drip-Sugar Cane Methods in the Aquaphonics System

Latif Sahubawa, Bambang Triyatmo and Erlina Ambarwati

Abstract

Aquaculture produces a large amount of wastewater, which can be controlled with an aquaponic systems that combines the cultivation of red tilapia and green mustard by converting the wastewater into a source of nutrients for biota and maximizing waste utilization in the fields of bioconversion and bioeconomics. The aim of the study was to analyze the growth of green mustard vegetable biomass and economic value in one cycle of red tilapia aquaculture. The physical-chemical quality of red tilapia aquaculture water bodies is pH = 7.4, temperature = 27.4°C, TDS = 184.3 mg/L, EC = 392 µs/cm, NH_3 = 6.4265 mg/L, NO_2 = 0.6204 mg/L, NO_3 = 4.178 mg/L. The physical-chemical quality of wastewater in green mustard cultivation is pH = 7.5, temperature = 27.6°C, TDS = 188.6 mg/L, EC = 399.7 µs/cm, NH_3 = 0.0025 mg/L, NO_2 = 2.6315 mg/L, NO_3 = 5.260 mg/L. The physical-chemical quality of wastewater for vegetable cultivation in vegetable pots is pH = 7.6, temperature = 27.6°C, TDS = 188.1 mg/L, EC = 400 µs/cm, NH_3 = 0.0026 mg/L, NO_2 = 2.8245 mg/L, NO_3 = 5.266 mg/L. Biomass growth of green mustard in the recirculation system = 92.50 kg per harvest and in drip system = 138.00 kg. The ratio of the growth of the top stem and rootstock of mustard greens is 4.0:1.0 and 5.5:1.0. The growth of red tilapia biomass at harvest reaches IDR 1,680,000. The total income from fish aquaculture and vegetable cultivation at each fish harvest period is IDR 10,323,750.

Keywords: aquaponics, bioconversion, bioeconomics, biomass, wastewater, red tilapia

1. Introduction

Indonesia's 2019–2023 National Medium-Term Development Plan outlines seven general policy directions, one of which aims to increase the added value of marine and fisheries resources. The Ministry of Marine Affairs and Fisheries of the Republic of Indonesia promotes a blue economy approach in national development planning. This approach focuses on the optimal use of fisheries industry waste to produce valuable food and non-food products, based on a zero waste. Waste from the aquaculture industry is rich in nutrients sourced from uneaten feed and feces (containing nitrogen in the form of total ammonium nitrogen and nitrite). When wastewater from aquaculture is discharged into the environment, these nutrients will be carried away, and cause pollution. The most commonly practiced and environmentally friendly aquaculture system in Indonesia is aquaponics [1].

Aquaponics-based aquaculture is an integrated approach that combines fish and vegetable cultivation to increase economic value by reusing wastewater as a source of nutrients for vegetable and fruit cultivation in supporting the production of quality organic food [1]. According to Ref. [2], intensive-scale aquaponics with high-quality feed has a significant impact on fish biomass growth and provides essential nutrients (nitrate), increasing the potential for integrated vegetable and fruit cultivation. In aquaponics system cultivation, nitrifying bacteria can convert nitrite to nitrate, which is the main component of liquid fertilizer nutrients in vegetable plants [1]. Using fish farming wastewater for vegetable cultivation is an effective solution to minimize pollution in fish pond water bodies, increase optimal fish growth, produce organic vegetables, save groundwater, and control environmental pollution [2]. In the era of modern technology, quality food can be produced independently and on a large scale. Aquaponics is a practical and environmentally friendly technology for cultivating productive and economically valuable plants on limited land while maintaining the sustainability of water resources [3, 4].

Advances in science, technology, and innovation in the field of aquaponics have made aquaculture wastewater optimally utilized as a source of nutrition in vegetable and fruit cultivation (horticulture). The application of the aquaponic cultivation system is theoretically, practically, and economically very profitable because the land used is relatively limited, multipurpose (fish, various vegetables, and fruits), water efficient, prevents the use of antibiotics, and is free from antibiotic residues. The technical-practical advantages are that there is no need to cultivate the land, free from fertilizers and pesticides, without irrigation and utilizing fish farming wastewater which is rich in nutrients. According to Refs. [5, 6] several types of fish that are cultivated have economic potential and bright market prospects, namely tilapia, carp, pomfret, patin fish, catfish, and several types of important vegetables, such as spinach, mustard greens, kale, lettuce, celery, tomatoes, chilies, and eggplant. The advantages of aquaponic technology include easy application, space-saving, free from pesticides and herbicides, no need for regular water changes, faster and more controlled growth of fish and vegetables, a productive alternative source of income, labor savings, a healthy (organic) food source, and environmentally friendly [3, 4].

Red tilapia is a freshwater fish known for its high protein and low-fat content. Every 100 grams of red tilapia meat contains 17.8% protein, fat (2.7%), water (77.8%), and minerals (1.2%) [7]. Green mustard is a low-calorie vegetable but rich in nutrients, beneficial for health, and widely cultivated throughout the world. Green mustard is often used for salads, soups, and as a garnish. With only about 13 calories per 100 grams, green mustard is an ideal food for those on a diet because it only contributes a few calories, and reduces fat, thus helping with weight loss. Green mustard also contains sulfur compounds, which can reduce the risk of cancer (breast, prostate, and lung cancer) by helping the body eliminate toxins and prevent cancer cells. The essential minerals (iron, magnesium, and calcium) it contain support bone growth and prevent osteoporosis. Vitamin A (beta and alpha carotene) in green mustard can improve eye health and prevent oxidative stress on the retina.

Other important nutrients include folate and vitamin B6, which are beneficial for heart health by lowering homocysteine levels (a risk marker for cardiovascular disease). Each cup of collard greens provides about 20% of the recommended daily intake of potassium, which is essential for heart health because it acts as a vasodilator, reducing blood vessel tension and lowering cardiovascular stress. Additionally, the vitamin C in collard greens boosts immune function, reduces stress, and stimulates the production of white blood cells to ward off chronic disease. Vitamin C also plays a key role in collagen synthesis, helping to maintain healthy and rejuvenating skin.

Based on the research results [2], the physicochemical quality of the water of red tilapia fish farming ponds for green mustard cultivation still meets the water quality requirements, but the pH and TDS values are approaching the highest limit. In this study, an aquaponic system was applied by cultivating red tilapia as a source of nutrients for the growth of green mustard with recirculation and molasses methods. The development of this aquaponic cultivation technology is expected to empower fish farming communities so that they are able to utilize land, fisheries, and environmental resources optimally to improve welfare. With the support of university researchers, farmer groups can independently improve aquaponic technology, in order to produce organic fish and vegetables that meet domestic market demand, as well as reduce waste, and contribute to environmental pollution control.

2. The research objectives

The main objectives of this research are (1) analysis of the physical-chemical quality of the water body for red tilapia fish farming as a source of organic liquid fertilizer for green mustard greens cultivation, (2) analysis of the physical–chemical quality of wastewater from red tilapia fish farming for vegetable cultivation, (3) analysis of the bioconversion of fish farming waste on vegetable biomass growth, and (4) bioeconomic analysis of fish and vegetable farming in an aquaculture system.

3. Materials and method

3.1 Materials and equipment

The raw materials for the research were red tilapia (2000 fish, L = 8–11 cm); green mustard seeds, and fish feed. The main equipment used is a rectangular fish cultivation tank (fiberglass) (L = 2.5 m, width/W = 1.5 m, height/H = 1.0 m, diameter/Ø = 2.0 m), a cylindrical fish cultivation tank (Ø = 1.5 m, W = 1.5 m, H = 1.0 m), a rectangular vegetable cultivation tank (fiberglass) (L = 1.2 m, W = 0.6 m, H = 0.40 m), a vegetable cultivation recirculation tray (conduit pipe: Ø = 3 inches, L = 2.0 m, W = 1.2 m, vegetable pot Ø = 6 cm), a vegetable cultivation tray with a sugarcane drip system (conduit pipe: Ø = 1/2 inch, L = 9 m, vegetable pot Ø = 22 cm), TDS meter and pH meter.

3.2 Research procedures

The stages of aquaponic system cultivation research are: (1) installation of an aquaponic house along with tanks and racks for fish and vegetable cultivation; (2) installation of a water network (conduit, pump, and clean and waste water reservoir); (3) installation of a wastewater recirculation system for vegetable cultivation (conduit with an open system, molasses, and an open tank); (4) installation of a three-tiered rack (vegetable cultivation with a molasses system); (5) filling clean water in the fish cultivation tank; (6) installation of an electricity network; (7) spreading red tilapia seeds and planting green mustard seeds in the vegetable cultivation installation; (8) providing fish feed and circulating waste water in vegetable cultivation; (9) measuring the physical-chemical quality of fish pond water as well as the water discharge and quality of fish cultivation wastewater; (10) measuring the growth of vegetable and fish biomass.

3.3 Research methods

Laboratory measurement of water quality samples at the Laboratory of the Center for Environmental Health Engineering and Disease Control, D.I. Yogyakarta, and direct measurement at the research location using a pH meter (pen type, pH 009-l and TDS meter type YL-TDS2-A). Research data processing using quantitative, qualitative, and descriptive methods [8]. Physicochemical parameters of fish farming water quality and fish and vegetable farming wastewater, namely: pH, temperature (°C), TDS (mg/L); electrical conductivity/EC (μs/cm), NH_3 (mg/L), NO_2 (mg/L), NO_3 (mg/L), and wastewater discharge (L/minute). Bioconversion parameters (growth of mustard greens and red tilapia) and bioeconomics (selling value of vegetables and fish) in one harvest period.

4. Results and discussion

4.1 The physical-chemical quality of water bodies for red tilapia aquaculture and wastewater for the cultivation of green mustard vegetables (recirculation and drip-sugar cane systems)

The physical-chemical quality parameters of pond water for red tilapia cultivation and red tilapia cultivation wastewater for green mustard vegetable cultivation are shown in **Table 1**.

No	Containers	Water and wastewater quality parameters						
		pH	Temperature (°C)	TDS (mg/L)	EC (μs/cm)	NH_3 (mg/L)	NO_2 (mg/L)	NO_3 (mg/L)
1	Fishpond of red tilapia	7.4	27.4	184.3	392	6.4265	0.6204	4.178
2	Paralon (recirculaion system)	7.5	27.6	188.5	399	0.0025	2.6315	5.260
3	Pot (drip-sugar cane system)	7.6	27.6	188.1	400	0.0026	2.8245	5.266

Table 1.
Physical-chemical quality of fishpond water bodies and wastewater for green mustard cultivation.

4.1.1 Acidity (pH)

The pH value of the water body of the red tilapia fish farming pond and wastewater for green mustard vegetable cultivation ranged from 7.4 to 7.6 (**Table 1**). The pH levels in the three cultivation containers were relatively the same, indicating that the pH of the fish aquaculture environment and wastewater used for vegetable cultivation remained neutral and stable for the growth of fish and vegetables. When the pH increases above 7.0 in aqueous solution, most of the dissolved phosphorus can react with calcium to form calcium phosphate. The high pH in the pot container can cause low nutrient absorption by the roots, thus inhibiting vegetable growth. According to Ref. [9], a lower pH level creates an optimal environment for growth with increased iron absorption so that the harvest period of vegetables is increased.

There are 6 (six) key factors that affect the growth of aquaponic and hydroponic plants, namely: water quality, sunlight, oxygen, nutrients, temperature, and pH of the nutrient solution. In the aquaponic system, 2 (two) important environmental parameters are pH and electrical conductivity (EC) because both have a significant impact on the physiology of aquatic life and cultivated plants [2]. Acidity affects the nutrient absorption capacity of plants and the activity of nitrifying bacteria (nitrite formation). Nitrifying bacteria convert ammonia from food waste and fish feces into nitrite, which is then converted by nitrobacter bacteria into nitrate at an optimum pH (±8.5). Optimal nutrient absorption by some plant species can occur at pH 6.0 [10]. The range of pH values in aquaponic system cultivation must be strictly controlled to achieve optimal conditions so that they can support the life and activity of nitrifying bacteria and the growth of cultivated vegetables. High acidity levels in water bodies can reduce the efficiency of absorption of essential macronutrients by plant roots in aquaponic systems.

The acidity level affects the growth and survival of aquaponic/hydroponic plants. Plants require a nutrient solution/water with a neutral pH (6.0–7.5) in order to grow and produce well and survive for a long time. Nutrients contained in solutions/water with neutral pH conditions (6.0–7.5) can be optimally absorbed by plant roots, while in acidic and alkaline pH conditions (6.0 ˂ pH ˃ 7.5) it turns out that it is not good for plant growth. In alkaline nutrient solutions/water, it turns out that micronutrients (Mn, Zn, Fe) will be chemically bound so that they cannot be optimally absorbed by plant roots, as a result, plants experience nutritional deficiencies and stunted growth. Only a small number of plants can grow well in pH conditions of 6.0–7.5 [11].

4.1.2 Temperature (°C)

The water body temperature of red tilapia fish farming ponds and green mustard cultivation wastewater ranges from 27.4 to 27.6°C (**Table 1**). Tilapia fish can grow normally at a temperature range of 14–38°C and spawn naturally at a temperature of 22–37°C. The optimal temperature for the growth of red tilapia fish ranges from 25 to 30°C [12]. The growth of red tilapia fish will be disrupted if the water body temperature is lower than 14°C and or higher than 38°C. At temperatures lower than 6°C or higher than 42°C, tilapia fish experience severe stress and die. This condition shows that the water body temperature of cultivation ponds and vegetable cultivation is still in the good category and optimal for growth.

One of the factors that directly affects fish survival is temperature. Fish can die if the water temperature decreases or increases beyond its tolerance limit. This is because fish are very susceptible to drastic changes in temperature [13, 14]. Many factors can affect temperature fluctuations, such as the temperature of the aquatic environment, dissolved particles, and sunlight, which are limiting factors in the growth of aquatic biota and plants [2]. The optimal environmental temperature will stimulate the growth of aquatic biota and plants optimally. The growth rate of aquatic biota and plants can take place optimally when the water temperature reaches its optimum because other environmental factors become neutral. When the water temperature changes from minimum to optimum conditions, the growth rate of aquatic biota and plants increases drastically at the highest limit [15]. According to Ref. [16], when dissolved particles are high, it will affect the ability to absorb sunlight. Low dissolved particles can cause the water temperature to decrease. The nitrification process requires a lot of energy, which will result in an increase in temperature. Changes in water body temperature at optimum to maximum intervals result in a decrease in the growth rate of aquatic biota/plants, except for certain types of biota and plants that grow quite quickly.

In vegetable cultivation in water media with moderate conditions, the optimum temperature for photosynthesis is lower than the optimum temperature for the respiration process. This is because the photosynthesis process will decrease in intensity if the temperature does not support it. The optimum temperature range for accelerating the photosynthesis process of aquatic plants ranges from 10 to 30°C. According to Ref. [5], if the environmental temperature is higher and/or lower than the optimum temperature conditions, the rate of photosynthesis will decrease, although there are several types of plants that are able to photosynthesize optimally. This shows that the temperature in the aquaponic system is important for the aquaculture ecosystem.

4.1.3 Total of dissolved solids (TDS, mg/L)

Based on the measurement results (**Table 1**), the TDS content in the water body of red tilapia fish cultivation was 184.3 mg/L and in vegetable cultivation with a recirculation system of 188.5 mg/L and 188.1 mg/L in the molasses system. The optimal TDS condition is less than 200.0 mg/L [2]. Treatment of vegetable cultivation with the recirculation and molasses methods can increase TDS. According to Ref. [17], the addition of nutrients to the aquaponic system brings benefits to plants and fish, but excessive addition of nutrients can affect TDS levels which can directly harm fish, as well as reduce pH and oxygen solubility in water. Therefore, the addition of nutrients should be limited to elements needed by plants in limited quantities.

TDS is an indicator of the number of organic and inorganic particles or solids in the form of calcium, magnesium, potassium, carbonate, nitrate, bicarbonate, chloride, and sulfate salts that are smaller than 1.0 nanometers [18]. TDS directly affects the turbidity and brightness of water bodies and the life of aquatic biota. Turbidity is a description of the optical properties of water due to the presence of suspended solids (clay particles, mud, soil colloids, and aquatic organisms). Suspended solids interfere with fish life and prevent sunlight from entering for phytoplankton photosynthesis. Turbidity is an important parameter of the quality of aquaculture water bodies. The dominant suspended material is formed from detritus, so that within certain limits it can be used as an indicator of water body pollution and has a significant effect on respiration, the ability to see aquatic biota, decreased dissolved oxygen levels, and inhibition of light penetration into water bodies.

Water clarity is a measure of the transparency of water and sunlight in a body of water. The intensity of sunlight for phytoplankton photosynthesis is influenced by the intensity and propagation power of sunlight and water turbidity. Water turbidity describes the optical properties determined by the amount of sunlight absorbed and emitted by dissolved and suspended materials in the water. The turbidity of the water body is influenced by fine suspended materials (mud), microorganisms (plankton), and the color of the water caused by colloidal substances resulting from the decay of plant leaves. The criteria for the clarity of fish pond water can be used as a reference in managing the quality of water bodies. Brownish pond water indicates the dominant growth of diatomaceous microalgae. Diatomaceous earth is one of the natural food suppliers for cultivated biota, so that its growth rate is relatively faster [5].

4.1.4 Electrical conductivity (EC, μs/cm)

One of the important parameters in aquaponic and hydroponic cultivation systems is electrical conductivity (EVC). EVC affects the physiological processes and yields of cultivated plants. The average EVC value in fish pond water bodies and vegetable cultivation is relatively the same, ranging from 392 to 400 μs/cm (see **Table 1**), but in vegetable cultivation water bodies it is relatively higher. The optimal EVC value is less than 450.0 μs/cm. EVC of water bodies significantly affects the ability of microalgae to absorb nutrients. This causes changes in stomatal openings and an increase in leaf area (photosynthesis efficiency). Several research results show that EVC values in aquaponic systems range from 300 to 1100 μs/m and in hydroponic systems between 1000 and 3000 μs/m. In aquaponic cultivation systems, the concentration of nutrients available in water is influenced by the type of fish, growth rate, and density of fish. In hydroponic system cultivation, the concentration of nutrients available in water is influenced by the type of fish, growth rate, and density of fish. The frequency of feeding and composition of feed have a significant effect on the growth rate of nitrifying bacteria [10].

The quality of nutrients in water bodies can be controlled based on EC and pH values. The higher the nutrient content in water bodies, the more concentrated the salt content in the water, so the EC capacity is higher (see **Table 1**). The increase in nutrient concentration in water bodies is influenced by the total salt content and accumulation of ions in water bodies. EC in water bodies affects plant metabolism, especially the rate of photosynthesis, enzyme activity, and the potential for ion absorption by plant roots. EC reflects the ion content in nutrients, where a decrease in EC values indicates the amount of ions absorbed by plant roots. Nutrients are formed from ions which are a mixture of macro and micronutrients needed by plants. The combination of these elements will have different EC values, directly proportional to the total dissolved ions [13, 14].

4.1.5 Ammonia (NH_3, mg/L)

Based on the measurement results (**Table 1**), it is known that the NH3 content in the water body of the red tilapia pond is 6.425 mg/L, higher than that of the green mustard cultivation container (0.025 and 0.026 mg/L). According to BSN, the ammonia content in fish cultivation is <0.02 mg/L, which indicates that the treatment of vegetable cultivation using a recirculation system and molasses is quite high. The difference in ammonia levels is caused by differences in the number, activity, and capacity of anaerobic bacteria (bacteria that decompose ammonia and nitrite) and aerobic bacteria (decompose nitrite into nitrate), and the process of water recirculation from the fish pond to the waste storage tank. The amount of fish feed given 4 – 5 times a day is the main factor triggering an increase in ammonia levels in the water body, compared to ammonia levels in vegetable cultivation containers. The low levels of ammonia in vegetable cultivation containers are caused by 2 (two) main factors, namely (1) the speed of decomposition of ammonia and nitrite by nitrifying bacteria in wastewater reservoirs (recirculation), and (2) the speed of absorption by vegetable roots [15].

The high ammonia levels in the water bodies of cultivation are an important indicator of the nutrient content of the water body which results in nutrient enrichment (eutrophication), followed by low dissolved oxygen levels, increased biological oxygen demand (BOD) and total suspended solid (TSS), and the emergence of fish pests and diseases (especially fungi and bacteria). This condition shows that if the accumulation of nutrients in fish ponds is not followed by drainage or suction and/or used every day as a source of nutrients for vegetable cultivation (aquaponic systems), it will reduce the quality of the water body to the lowest limit that affects fish life. The increase in ammonia and nitrite levels in large amounts becomes a toxic substance in the water body so it can cause mass fish deaths. The levels of ammonia and nitrite in the water body of fish ponds must be strictly controlled through a continuous aeration process to create aerobic conditions in the water body to accelerate the growth of nitrobacteric bacteria that decompose nitrite into nitrate and nitrifying bacteria that accelerate the decomposition of ammonia and nitrate compounds into nitrate (nutrients for aquatic plants) [19].

4.1.6 Nitrite (NO_2, mg/L)

Based on the measurement results (**Table 1**), it is known that the nitrite level in the red tilapia fish cultivation water body is 0.6204 mg/L, smaller and significantly different from the wastewater used for vegetable cultivation with a recirculation system and molasses (2.6315 – 2.8245 mg/L). This very striking difference in NO_2 levels indicates that in the red tilapia fish cultivation pond, optimal NO_2 decomposition occurs by nitrobacter bacteria, which is supported by the availability of oxygen in sufficient quantities from the aerator. In contrast, high NO_2 levels in the green mustard cultivation container are due to accumulation symptoms. Similar to ammonia, nitrite levels in the water body of the fish pond must be strictly controlled through appropriate conditioning to accelerate the growth of nitrifying bacteria whose job is to decompose ammonia into nitrite (a poison for cultivated biota) and nitrobacter bacteria that accelerate the decomposition of NO_2 into NO_3 [20].

Nitrogen comes from organic waste in various forms (organic N, ammonia N, nitrite N, and nitrate N). In industrial wastewater, the organic N content is relatively higher than ammonia N, while in wastewater with warm temperatures, the organic N content is relatively lower than ammonia N. Nitrite and nitrate in wastewater are at very low concentrations [20]. Nitrogen forms undergo transformation as part of the nitrogen cycle, namely: (1) assimilation of inorganic N (NH_3 and NO_3) by plants and autotrophic bacteria to form organic nitrogen (amino acids and proteins); (2) oxidation of NH_3 to NO_2 and NO_3 by aerobic bacteria occurs optimally at pH 8 in water body temperature conditions of ±30°C; (3) ammonification of organic N produces NH_3 during the decomposition process of organic matter, mostly carried out by microbes and fungi; (4) reduction of nitrate to nitrite, nitrogen oxide (N_2O), and nitrogen (N_2) occurs optimally in anoxic conditions involving bacteria and fungi.

4.1.7 Nitrate (NO_3, mg/L)

Based on the results of the analysis (**Table 1**), nitrate levels in red tilapia aquaculture ponds are 4.178 mg/L, smaller and significantly different compared to recirculation vegetable cultivation containers (5.260 mg/L) and drip-sugar cane systems (5.266 mg/L). This condition shows that the activity of nitrobacter bacteria in the decomposition of nitrite into nitrate is higher in the reservoir of redfish aquaculture because it is supported by conditions of temperature and pH and the availability of oxygen. The higher pH of the water body in the fish culture pond affects the higher ammonia toxicity, because most of it is in the form of NH_3, while ammonia in the molecule (NH_3) is more toxic than in ionic form (NH_4^+). Ammonia in the form of molecules can form cell membranes faster than NH_4^+ [21]. According to Ref. [22], the source of ammonia in fishpond water comes from fish feed and fish feces. The amount of ammonia produced can be estimated from the use of net protein (added protein) from the feed given.

According to Ref. [23], chemical compounds of ammonia nitrogen (N-ammonia) in water bodies sourced from food and fish feces provide growth stimulation in aquatic plants. The concentration of organic and inorganic nitrogen compounds in the waters will change if supported by environmental conditions so that it will cause new problems in water bodies. According to Ref. [21], the concentration of organic nitrogen in water bodies varies greatly, depending on fish cultivation technology. Ammonia nitrogen (NH_3-N) concentrations are quite high in ponds that are given fertilizer rather than only biber feed. Nitrogen also contains dissolved organic matter ±1.0 mg/L in unpolluted waters, while in waters with high nutrition (2–3 mg/L) can cause phytoplankton blooms.

4.2 Discharge, bioconversion, and bioeconomics of wastewater for green mustard cultivation

4.2.1 Wastewater discharge capacity for the cultivation of green mustard vegetables

Based on the data in **Table 2**, it can be seen that the discharge of red tilapia fish cultivation wastewater for green mustard cultivation with a recirculation system is greater than the discharge of vegetable cultivation wastewater using a molasses system, namely: (1) discharge of red tilapia fish cultivation water for green mustard cultivation with a recirculation system = 6.3–6.6 L/minute (2 PVC trays with 126 vegetable pots, Ø = 8 cm), and (2) discharge of red tilapia fish cultivation water for green mustard cultivation with a molasses system = 5.2–5.8 L/minute (1 iron rack, stacked 3, with 36 vegetable pots, Ø = 22 cm).

The aquaponics system has three fish farming tanks, one freshwater reservoir, and one fish farming wastewater reservoir with three vegetable farming models (two PVC trays for the recirculation system, six fiberglass tanks for the floating system, and one iron rack for the molasses system) (**Figure 1**). Wastewater from red tilapia farming activities is available in large quantities (capacity 0.72 m^3) which is continuously flowed to each vegetable farming tray. Based on daily observations, the available wastewater can be used continuously to meet the needs of large-scale economic vegetable cultivation (leaf vegetables: green and white mustard greens, green and red spinach, kale, green and red lettuce, and fruit vegetables: chili, tomato, eggplant, red onion) and fruits.

The main obstacles in aquaponic cultivation systems that must be anticipated to increase cultivation productivity are: (1) reducing the total suspended solids that can block the piping in the molasses-sugarcane cultivation system and plant roots; (2) designing a wastewater reservoir that has a special column at the bottom to settle suspended particles so that they do not enter the recirculation channels to the vegetable

No	Product type	Initial weight (g)	Harvest (per season)		Price sell (IDR/kg)	Harvest period (times)	Amount (IDR)
			Total length (cm)	Total weight (kg)			
	(1)	(2)	(3)	(4)	(5)	(6)	(7) = (4) × (5) × (6)
1	Production of green mustard						
	Paralon (recirculation system, A)	4.5–6.5	30.0–35.0	92.50	7500	1	693,750.00
	Pot (molasses system, B)	4.5–6.5	35.0–42.0	138.00	7500	1	1,035,000.00
	Total—1	—	—	230.50	7500	5	8,643,750.00
	Ratio of biomass (A/B)	—	—	1.0: 1.5	—	—	—
2	Production of red tilapia	8–10	11–22	48.00	35,000	1	1,680,000.00
	Total—2	—	—	—	—	—	1,680,000.00
	Total 1 + 2	—	—	—	—	—	**10,323,750.00**
3	Economic ratio of green mustard and red tilapia	—	—	—	—	—	5.14:1.00
4	Wastewater discharge (L/minute)						
	Paralon (recirculation system): 6.3–6.6 L/minute						
	Pot (molasses system): 5.2–5.8 L/minute						

Notes: A = cultivation of green mustard vegetables with a recirculation system; B = cultivation of green mustard vegetables with a molasses system.

Table 2.
Bioconversion and bioeconomics analysis of aquaponic systems (fish and vegetables).

cultivation unit; (3) recirculation of wastewater in vegetable cultivation tanks must be carried out for a relatively long time and repeatedly before being distributed to the pond so that the nutrients can be maximally absorbed by the vegetable roots; and (4) maximizing the use of nutrient-rich aquaculture wastewater for the cultivation of various economical vegetables and fruits with an aquaponic system (vertical) and on narrow land [9, 16].

4.2.2 Bioconversion and bioeconomic analysis in aquaponic system cultivation

Bioconversion and bioeconomic analysis on the utilization of aquaculture wastewater for green mustard cultivation are described in **Table 2**. From the results of the bioconversion analysis, it is known that the total growth of green mustard biomass with recirculation system cultivation = 92.50 kg and molasses system = 138.00 kg (ratio 1.0:1.5) with total production = 230.50 kg and red tilapia

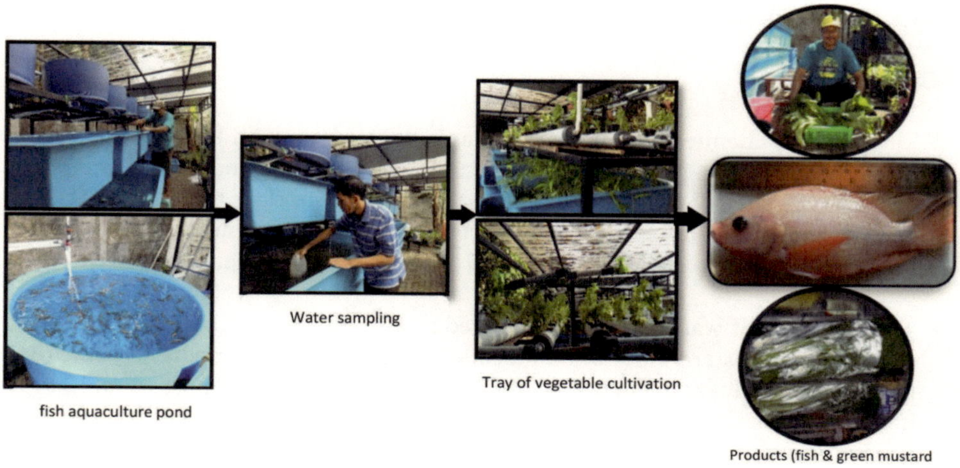

Figure 1.
Model of aquaponic system aquaculture (integrated of fish and vegetable cultivation).

production = 48.00 kg. The results of the bioeconomic analysis show that the selling value of green mustard vegetables is IDR 8,643,750.00 and red tilapia is IDR 1,680,000.00 (Total = IDR 10,323,750.00). Based on these results, it is known that vegetable cultivation with the molasses system is more productive than the recirculation system.

Aquaponic system cultivation is one of the smart cultivation models in an effort to increase the income of fish and vegetable farmer groups, by utilizing narrow land, without using water and large space. Aquaponic system cultivation is very suitable to be developed in marginal lands with limited groundwater but has high economic potential [17, 24]. On the other hand, the fish and vegetables produced have great market potential with quite competitive prices in traditional and modern markets.

4.2.3 The advantages of aquaponic system cultivation

Aquaponic cultivation system business is called "an alternative solution to solving environmental problems into an economic source," with the following advantages: (1) preventing environmental pollution by utilizing wastewater as liquid organic fertilizer in organic vegetable and fruit cultivation businesses that have economic value; (2) utilizing land and water resources optimally through vegetable cultivation (saving space) and reusing fish farming wastewater as an effort to conserve and rehabilitate groundwater; (3) aquaponic cultivation system technology is very easy to implement with relatively low operational costs and high output; (4) aquaponic cultivation system business saves a lot of labor because the operational system is easy to operate, can be operated by one or two people per business unit (categorized as a very economical business); and (5) vegetable cultivation in the aquaponic aquaculture system can be harvested in a relatively short time (around 25 days) so that it has a greater economic value than fish cultivation (main product). Aquaponic plant cultivation business (green mustard greens and red tilapia) with a recirculation system and molasses, has a greater economic value ratio for vegetables than red tilapia, reaching 5.14:1.0.

5. Conclusions and recommendations

5.1 Conclusions

1. The physical-chemical quality of red tilapia aquaculture pond water still meets the water quality requirements for the cultivation of green mustard vegetables even though the pH value and total dissolved solids are close to the highest limit. The aquaponic technology used in the cultivation of green mustard vegetables produces fairly stable water quality and high biomass productivity.

2. The discharge of wastewater from the red tilapia aquaculture business for vegetable green mustard cultivation reached 5.2–6.6 L/minute, apparently able to provide a significant influence on the growth of vegetable biomass (230.00 kg) in the recirculation and drip-sugar cane systems in 5 (five) times of harvest and production of red tilapia of 48 kg. It shown that the discharge of wastewater from red tilapia aquaculture for the cultivation of green mustard greens with a recirculation system is greater than the discharge of wastewater for vegetable cultivation of the drip-sugarcane system.

3. The total economic value of the production of red tilapia and mustard greens in one cultivation cycle reaches IDR 10,323,750.00. These results demonstrate the efficiency and potential for the sustainability of this aquaponic system.

5.2 Recommendations

1. The use of red tilapia aquaculture water for vegetable cultivation businesses should be given a sufficient waiting time (about 1 day) before use, in order to excite and convert ammonia and nitrite to nitrate to the maximum while reducing the pH of wastewater close to neutral conditions (pH < 7.5).

2. The discharge of aquaculture wastewater is large enough so that it must be utilized optimally for the development of a variety of important economic and fruit cultivation businesses on a larger scale.

Acknowledgements

There are countless thanks and awards conveyed to the honorable: Director of Research and Community Service, Kemristekdikti Jakarta, Chairperson of UGM LPPM and Dean of the Faculty of Agriculture UGM who have given us approval and trust to carry out this research for three fiscal years (2017, 2018, 2019).

Author details

Latif Sahubawa[1*], Bambang Triyatmo[1] and Erlina Ambarwati[2]

1 Department of Fisheries, Faculty of Agriculture, Gadjah Mada University, Yogyakarta, Indonesia

2 Department of Agriculture Cultivation, Faculty of Agriculture, Gadjah Mada University, Yogyakarta, Indonesia

*Address all correspondence to: lsahubawa@ugm.ac.id

References

[1] Anonimous. Presidential Regulation of the Republic of Indonesia, Number 2, 2015 concerning National Medium Term Development Plan (RPJMN), Indonesia 2015-2019. Indonesia: Ministry of National Development Planning; 2015

[2] Sahubawa L, Triyatmo B, Ambarwati E. Bioconversion and bioeconomic of wastewater from Red Tilapia aquaculture on the aquaponics system as source of nutrient in green mustard growth. In: E3S Web of Conferences. EDP Sciences; 2020. p. 01013

[3] Junus SL. Design of Tray for Cultivating Environmentally Friendly Aquaponic Fish Systems. Summary of Results of Discussion of UGM Akuaponik Research Team with Partners (KUB Mina Abadhi Farm Yogyakarta), Yogyakarta (unpublished). 2017

[4] Wagiman. Practical Module for Industrial Waste Control. Study Program of Strata I. Department of Agricultural Industrial Technology. Yogyakarta: Faculty of Agriculture Technology, Gadjah Mada University; 2014

[5] Roidah IS. Land use by using hydroponic systems. Journal of Tulungagung University, Bonorowo. 2014;**1**(2):43-50

[6] Sahubawa L, Triyatmo B, Ambarwati E. Economic and institutional strengthening of aquaculture micro-small scale aquafonics systems. In: Paper Presented at the National Seminar on Marine and Fisheries Socio-Economics. The Center for Marine and Fisheries Socio-Economic Research (BBRSEKP) Collaborates with the Indonesian Marine and Fisheries Socio-Economics Research Network (IMFISERN) and the Conservation Strategy Fund (CSF) Indonesia. Mina Bahari IV Building, Jakarta Ministry of Maritime Affairs and Fisheries, Jalan Batu No. 1, Gambir - Jakarta Pusat, Septeber 24, 2018. 2018

[7] Nurlaeny N. Teknologi Media Tanam dan Sistem Hidroponik. Bandung: Unpad Press; 2014

[8] Sugiyono. Metode Penelitian Pendidikan Pendekatan Kuantitatif, kualitatif, dan R&D. Bandung: Alfabeta; 2010

[9] Wang YJ, Yang T, Kim HJ. pH dynamics in aquaponic systems: Implications for plant and fish crop productivity and yield. Sustainability. 2023;**15**(9):7137

[10] Nugroho CA, Setiawan AW. Pengaruh frekuensi penyiraman dan volume air terhadap pertumbuhan tanaman sawi pakcoy pada media tanam campuran arang sekam dan pupuk kandang. AGRIUM: Jurnal Ilmu Pertanian. 2022;**25.1**:12-23

[11] Pantella R, Rosta, and Hamidpour. Crop physiological response to nutrient solution for electrical conductivity and pH in an ebb-and-flow hydroponic system. Scientia Horticulturae. 2012;**194**(1):34-42

[12] Parsono S, Zakiyuddin A, Utami I. Penerapan Teknologi Media Tanam Hidroponik untuk Mewujudkan Wisata Pertanian di Desa Cimekar. Jurnal Pengabdian Tri Bhakti. 2021;**3.1**:14-22

[13] Domingues R, Barbosa AB, Galfawo H. Ecological Indicators. 2012;**24**:245-255. DOI: 10.1016/j.ecolind.2012.06.020

[14] Brito AC et al. Replying to Domingues et al., Ecological Indicators. Ecological Indicators. 2013;**27**:123-124. 24, 245-255. DOI: 10.1016/j.ecolind.2012.06.020

[15] Khairuman, Amri. Introduction to Morphology and Anatomy of Fish Reproduction. Practical Material of Reproduction Physiology in Aquatic Animals. Department of Aquaculture, Faculty of Agriculture, University of Lampung; 2007

[16] Prasetiawan H, Mukadis A, Motti AV, Nurhayati, and B. Setiawan. Study of nitrification process in a media raised bed based aquaponic system. Journal of Clean Technology. 2024;**01**(2):82-89

[17] Hutagalung RA, Seowono AD, Darmawan M, dan A. Cornelius. Effects of nutritional and culture medium-based approaches for aquaponics systems with bio-floc technology on Pak Choi and catfish growth rates. Tropical Agriculture Science. 2023;**46**(4):1259-1274

[18] Megayanti L, Zurhalena Z, Junedi H, Fuadi NA. Kajian beberapa sifat fisika tanah yang ditanami kelapa sawit pada umur dan kelerengan yang berbeda. Jurnal Tanah Dan Sumberdaya Lahan. 2022;**9.2**:413-420

[19] Siregar A. Content and Impact of Ammonia, Nitrite and Nitrate in Industrial Wastewater. 2005. Available from: http://www.blogspot.com [Accessed: November 19, 2018]

[20] Prasatya. Feasibility Analysis and Optimization of Cultivation of Green Mustard and Hydroponic Vegetable Cultivation with the NFT System (Nutrient Film Technique) at PT. Joy Farm, Depok: Final Project, Faculty of Agricultural Technology, Bogor Agricultural University; 2009

[21] Sari W, Kitty Y, Dwiranti A. Hydroponic system with the nutrient film technique (NFT) and Wick on planting green mustards. Surya Octagon Interdisciplinary Journal of Technology. 2016;**1**(2):223-225. ISSN: 2460-8777

[22] Kordi, Andi. Water Quality Parameters for Fish Cultivation. Fisheries Extension, Fish Cultivation, Fish Processing and Ornamental Fish. Jakarta: Textbook, Publisher PT. Penebar Swadaya Nelayan Abadi; 2009

[23] Nur FF, Sirajuddin H, Abdullah, and P. Asih. Water quality analysis in aquaponic irrigation systems. Analysis of water quality in the aquaponic irrigation system. Scientific Journal of Agricultural Engineering and Biosystems. 2017;**5**(2):385-394

[24] Prayogo P, Agustono A, Rahardja BS, Amin M. Growth performance and nutrient composition of mustard green (Brassica juncea) cultured in aquaponics systems and hydroponic system. Journal of Aquaculture and Fish Health. 2021;**10.3**:373-379